THE GOLD RUSH LETTERS OF

E. ALLEN GROSH AND

HOSEA B. GROSH

WILBUR S. SHEPPERSON SERIES IN NEVADA HISTORY

THE

Gold Rush Letters

OF **E. ALLEN GROSH** AND
HOSEA B. GROSH

EDITED BY

Ronald M. James and
Robert E. Stewart

UNIVERSITY OF NEVADA PRESS RENO AND LAS VEGAS

WILBUR S. SHEPPERSON SERIES IN NEVADA HISTORY

This publication is made possible in part by a grant from the Nevada Humanities, a state program of the National Endowment for the Humanities.

University of Nevada Press, Reno, Nevada 89557 USA
www.unpress.nevada.edu
Copyright © 2012 by University of Nevada Press
All rights reserved
Manufactured in the United States of America
Design by Kathleen Szawiola

LIBRARY OF CONGRESS CATALOGING-IN-PUBLICATION DATA
Grosh, E. Allen, 1824–1857.
The gold rush letters of E. Allen Grosh and Hosea B. Grosh /
edited by Ronald M. James and Robert E. Stewart.
pages cm. — (Wilbur S. Shepperson series in Nevada history)
Includes bibliographical references and index.
ISBN 978-0-87417-885-2 (cloth : alk. paper) —
ISBN 978-1-943859-01-6 (paper : alk. paper) —
ISBN 978-0-87417-892-0 (ebook)
1. Grosh, E. Allen, 1824–1857—Correspondence. 2. Grosh, Hosea B., 1826–
1857—Correspondence. 3. Gold miners—California—Correspondence.
4. Gold miners—Nevada—Correspondence. 5. Comstock Lode (Nev.)—
History—19th century. 6. West (U.S.)—Gold discoveries.
I. James, Ronald M. (Ronald Michael), 1955– , editor
II. Stewart, Robert E. (Robert Earl), 1936– , editor III. Title.
F865.G76 2012
979.3′010922—dc23   2012012188

The paper used in this book meets the requirements of
American National Standard for Information Sciences—
Permanence of Paper for Printed Library
Materials, ANSI/NISO Z39.48-1992 (R2002).

This book has been reproduced as a digital reprint.

Frontispiece: Natives of Pennsylvania Hosea Ballou (*left*) and
Ethan Allen Grosh (*right*) traveled to the California gold country in 1849
while in their early twenties. *Courtesy Nevada Historical Society*

CONTENTS

ILLUSTRATIONS

FOREWORD

This volume presents an important collection of letters from the era of the California Gold Rush. It is an essential contribution for anyone interested in how this period changed the nation, and the Nevada Historical Society is honored to participate in its publication. The stirring story of Ethan Allen and Hosea Ballou Grosh, participants in one of the great events of world history, continues to captivate. The letters the young men left behind assume an indispensable place in western studies.

The Grosh brothers collection encompasses more than eighty signed letters, related correspondence from family and friends, original transcripts, and several additional images dating from 1849 through 1880. The letters, dating from 1849 to 1857, are the most important known correspondence in Nevada's preterritorial history. These documents detail the everyday lives of '49ers mining in California and Nevada, including the daily hardships, depression balanced against optimism, and illnesses endured by the brothers.

The Grosh and Wegman families kept the surviving documents and photographs that told the story of the Grosh brothers' unfulfilled mining claim through court cases and house fires. I would like to thank Fred N. Holabird, local historical consultant, for first introducing the society and the Wegman family, allowing talks about acquiring the highly significant letters of the Grosh brothers to begin in 1997. Manuscript curator Eric Moody took the lead in the society's fund-raising effort, seeking funds from supporters, grants, private foundations, and the Nevada Legislature and fulfilling his quest in 2007 after ten years of effort. The cooperation of Charles T. Wegman and his mother, Naomi Thompson, descendants of the Grosh family, was critical to the effort to place this treasure in the public domain, and they have the thanks of the Nevada Historical Society.

The idea of publishing this book began in earnest in 2007. Eric Moody and Mella Harmon, both members of the Nevada Historical Society, began talking with the University of Nevada Press about publishing the book and enlisted the help of Ron James, the Nevada state historic preservation officer, to write the introduction. With the change in staff at the society starting in July 2009, roles in relation to the publication changed, with Ron James and Bob Stewart taking the lead as coeditors. The original transcript of the letters completed by Charles Wegman in 1997 provided a great start for the manuscript. However, pertinent information was missing, including some text as well as context and notes that would help the reader understand the many topics the Grosh brothers discussed in the beautifully written and dutifully descriptive letters they wrote to their family.

As one of the promises made to the family, due to the fragile condition of the letters, the staff at the society began scanning the documents for preservation purposes to reduce handling and prevent the loss of information as the letters continue to crumble. Today, researchers can study the researchers' copy of the transcript and printed copies of the historic letters for comparison in the society's research library.

The Nevada Historical Society has the honor of being the oldest cultural institution in the state of Nevada. It was founded by Dr. Jeanne Elizabeth Weir, history professor at the University of Nevada, along with several members of the Nevada Academy of Sciences, on May 31, 1904. Our mission is to collect and preserve objects and to educate the public on the history of the state of Nevada, the Great Basin, and the West. The remarkable collections housed at the society for more than a century will continue to be a valuable resource for future generations.

The society's academic connection began with Dr. Weir's association as professor and director. This close affiliation has continued to the present, encouraged by the society's current location at the north end of the University of Nevada, Reno, campus on North Virginia Street. The Nevada Historical Society thanks the University of Nevada Press for its assistance in publishing this volume, the result of a partnership as both institutions seek to serve those who cherish the history of Nevada and the West.

Sheryln L. Hayes-Zorn
Nevada Historical Society

PREFACE

The California Gold Rush of 1849 was one of the great events of human history. Much has been written about the tens of thousands who arrived, and letters and diaries left by the Argonauts, as they called themselves, have proven invaluable as people try to imagine what it was like to arrive in California, itself transforming from the northern frontier of Mexico into a new territory of the United States of America. The Gold Rush captured the world's imagination just as it defined a region, making the West a place where people could dream of opportunities that seemed out of reach elsewhere. Perhaps too often, California's gold country in the ravines of the Sierra Nevada represented hard work and disappointment. But for a generation of young adventurers, those who were there could at least say with some satisfaction that they had been at the center of the excitement.

With the publication of this volume, the letters of two young men from Pennsylvania, Ethan Allen and Hosea Ballou Grosh, can take their place among the sources of insight into this dynamic time. The brothers were eloquent, careful observers of their new home, and they were prolific, providing details of their trek across Mexico, their life in San Francisco, and their repeated efforts to strike it rich on the western Sierra slope. Finally, they describe three forays into the Great Basin, where Allen and Hosea prospected in the vicinity of what would come to be known as the Comstock Mining District in modern-day Nevada. Regardless of this last venture, which has too often directed the historian's eye away from the other letters, the broad sweep of the documents allows consideration of an entire region as it took shape. Nearly a decade of correspondence offers a wide range of insights, combined with a literary quality not often encountered in primary sources.

Born in 1824 and 1826, the brothers died within a few weeks of one another in 1857. The final chapter in this nineteenth-century adventure occurred while they were prospecting in Gold Canyon, which meanders south of what was then an empty mountainside but would in two years become the location of Gold Hill and its more famous neighbor, Virginia City. Hosea, the younger of the two, struck his foot with a pickax. The wound festered, and he died of the infection. Several weeks later, after writing his father about Hosea's death, Allen tried to cross the Sierra. A November snowstorm left Allen and a traveling companion with crippling frostbite, which cost the surviving brother his life, but, again, not before writing his father a letter describing the disaster.

BECAUSE THE LETTERS ARE FROM ANOTHER TIME, their meaning can sometimes be lost on the modern reader. We offer notes when references that once made sense have drifted into obscurity or when names warrant further explanation. In addition, three topics called for more complete discussions, which appear in the appendixes. The first of these explains the Reading California Association, the financial cooperative that the Grosh brothers joined in Reading, Pennsylvania, throwing their lot in with others with the joint purpose of shared resources and profit. Like most of these ventures, the business failed almost as soon as the journey began, but the people involved and the complications of the contract that was supposed to bind the participants together dominate the earliest letters, warranting some context, which appears in appendix A.

Appendix B discusses an odd assertion that Allen and Hosea were daguerreotype artists. Although the letters never mention the brothers practicing this form of early photography, it seemed to us that this volume would be lacking if the subject were not discussed, particularly since authorities on daguerreotypes frequently list the young men as practitioners. Here, then, is an assessment of this subject.

The letters the young men sent home during their eight-and-a-half-year sojourn to the West were both sentimental keepsakes and evidence. With that second role in mind, their Pennsylvania family looked to the correspondence to bolster the argument that they were entitled to a share of the Comstock Lode, one of the greatest gold and silver deposits ever found. Perhaps the wish to preserve the words of the lost brothers might have sufficed to inspire the preservation of the letters, but the motivation of millions of dollars would certainly have enhanced the feeling that the documents should be kept. The validity of the Grosh-family legal argument for ownership of the Comstock is the subject of appendix C; what matters for the rest of this volume is the effect of a family's efforts to safeguard what has survived as one of the best collections of Gold Rush–era correspondence to see publication in recent years.

THE GROSH LETTERS range from the intact to the disintegrating. Having survived house fires between the 1850s and the present, many were scorched, leaving words on some edges and folds nearly obliterated. Tears and aging reduced legibility in other places. Because Grosh-family descendants transcribed the letters before the Nevada Historical Society acquired them in

2007, the typewritten text that accompanied the letters records some words that can no longer be seen in the documents. The versions printed here draw on this source of information.

Similarly, the Reverend Aaron B. Grosh quoted from two of Allen's last letters as part of the obituaries published for his sons in 1857 and 1858. Aaron changed the text for print, sometimes substituting what he apparently felt were better words. As a result, his transcriptions are suspect. They are useful, however, when considering gaps that later appeared in the letters because of deterioration over the century and a half that followed his readings. Where words are no longer legible, the father's transcripts have served to amend the text. Similarly, Eliot Lord published excerpts from the letters in his 1883 *Comstock Mining and Miners* and in an 1895 article for the *New England Magazine*. We checked his publications to confirm passages and to fill in gaps.

Where an omission could not be bridged by any defensible means, the original transcribers inserted an ellipsis, and we have also employed this approach. On occasion, we felt comfortable augmenting the text because part of a word survived or because the context allowed for a reasonable determination of what was missing. Besides this sort of issue, there were times when the writers of the letters, in their haste, skipped a word needed to make the meaning of the sentence clear. In the worst of these cases, we inserted the missing word or words in brackets. When parentheses appear in the letters, however, they are from the original.

On occasion, the combination of gaps in the letters and isolated words that survived resulted in text that conveys no reasonable meaning. Because this encumbers the reading of the letters, sequences of unintelligible words have been deleted, and the missing sections are indicated with single ellipses. We have provided the Nevada Historical Society with a complete manuscript, showing all the words that were believed to exist in the letters either through observation of the letters or through the examination of available previous transcripts. That version also offers clear indications of words we have surmised to exist in the original letters. This material is available for researchers who wish to delve further into the text of the letters.

Italics indicate the names of ships, newspapers, and book titles, for example, but since the brothers were writing in cursive, they could not use this distinction. Instead, they inconsistently put newspapers and ships in quotes, but we have chosen italics to mark these types of names. Underlining indicates that emphasis in the original letters.

Misspellings and inconsistent capitalizations are common in the letters. In addition, the brothers often employed British spelling conventions, and they frequently used abbreviations. In general, we avoided abbreviations and made spelling and capitalization consistent. It seemed to us that these sorts of variations do not convey sufficient information to justify presenting the reader with such obstacles, although some of the brothers' more imaginative approaches justified an explanatory note. An exception to this is Hosea's letter of June 13, 1849, from Mexico, when he attempted to spell "diarrhea" as "diarrha" and "diahrhea." He concluded, "I declare I can't spell that." Preserving the incorrect spellings of the word was critical to understanding the letter, so they were presented as written. It is also appropriate to note that the brothers later spelled the word correctly. Perhaps because they were often afflicted, they had reason to become all too familiar with the term.

Punctuation in the letters is sometimes lacking to such a degree that it is difficult to understand the author's intent. Hosea, in particular, neglected the use of commas and periods. We found it necessary to add punctuation on occasion to clarify meaning. At the same time, while still more punctuation was arguably appropriate, only a minimum was inserted or modified to preserve the flavor of the original text. As always, researchers are encouraged to consult the actual documents to understand the full spectrum of information these valuable letters offer.

MORE THAN A CENTURY AND A HALF have compromised the letters, but we—and historians of the West in general—owe a debt of gratitude to the Grosh family for preserving these documents. This is all the more the case when acknowledging family members, and in particular Charles T. Wegman, for transcribing the letters.

In addition, we wish to thank the many other people who contributed to this volume's production. As described in the foreword, staff at the Nevada Historical Society played pivotal roles in imagining what form the publication would assume. Specifically, we would like to thank the society's Sheryln Hayes-Zorn, acting director, as well as Lee Brumbaugh, curator of photography, who helped a great deal with the scanning of the letters.

Regarding raising the funds to acquire the letters—in addition to those mentioned in the foreword—Bob Stoldal and Bob Ostrovsky, chairman and vice chairman, respectively, of the Advisory Board for Museums and History, played particularly important roles in soliciting support to acquire this material. Our special thanks go to Senator Bob Coffin of the Nevada Legislature for his assistance in this regard. In addition, private donors included Lee Mortensen, the Charles H. Stout Foundation, the Nevada History of Medicine Foundation, the Geological Society of Nevada Foundation, and Marilyn (Lynn) Bremer through the Mary Bremer Foundation. And there were smaller donations from a variety of other sources.

The following people assisted us with various aspects of the development of notes or other parts of this volume: Rafael Morales Bocardo, director, Historical Archive of the State of San Luis Potosí, Mexico; Michael Brodhead, emeritus professor, History Department, University of Nevada, Reno, and historian with the U.S. Army Corps of Engineers; William Chrystal, retired minister of the First Congregational Church, Reno, Nevada; Karen Dau, archivist, New York State Convention of Universalists; Robert and Marion Ellison, early Carson County, Utah Territory, historians; Katherine Fowler, emeritus professor, Anthropology Department, University of Nevada, Reno; Cathy Goudy, Marin County, California, Genealogical Society; Michael Green, Department of History, College of Southern Nevada; Eugene M. Hattori, curator of anthropology, Nevada State Museum; Fred Holabird, Holabird-Kagin Americana; James Hulse, emeritus professor, History Department, University of Nevada, Reno; Thomas R. Kailbourn, for information on daguerreian photography; Bruce Kirby, Manuscripts Division, Library of Congress; Michael J. Makley, historian of the Lake Tahoe Basin; Roger McGrath, University of California, Los Angeles, History Department; Anna Paredes, for Spanish translation; Thomas Perez, for image and location of the E. Allen Grosh grave at Last Chance; Wayne Reynolds, past international grand master of the Independent Order of Odd Fellows; George Rugg, curator, Department of Special Collections, University of Notre Dame; Cathy Reiter Scherer, Taylor family genealogist; Larry Schmidt, U.S. Forest Service, retired, and Nevada-California Trails historian; Sue Silver, El Dorado County historian; and Joseph Tingley, Nevada Bureau of Mines and Geology, University of Nevada, Reno. In addition, we would like to thank Amanda Leigh Brozana, National Grange communications director, for use of the image of Aaron Grosh.

We are also grateful to the staff of the University of Nevada Press. Matt Becker, acquisitions editor, played a particularly important role in shepherding this project through the process of development. And the support of Joanne O'Hare, director, is always appreciated.

Finally, we wish to express gratitude to those who endured the most, too often without recognition or proper thanks. Ron James thanks his mother, Wilma, his son, Reed, and his wife, Susan, for reviewing the manuscript and for their ongoing support. Bob Stewart thanks his wife, Phyllis, for enduring the long hours spent in research and for critically reviewing the notes and for her comments on other sections of the manuscript for clarity and readability. None of our work would have been possible without the encouragement of our families.

THE GOLD RUSH LETTERS OF

E. ALLEN GROSH AND

HOSEA B. GROSH

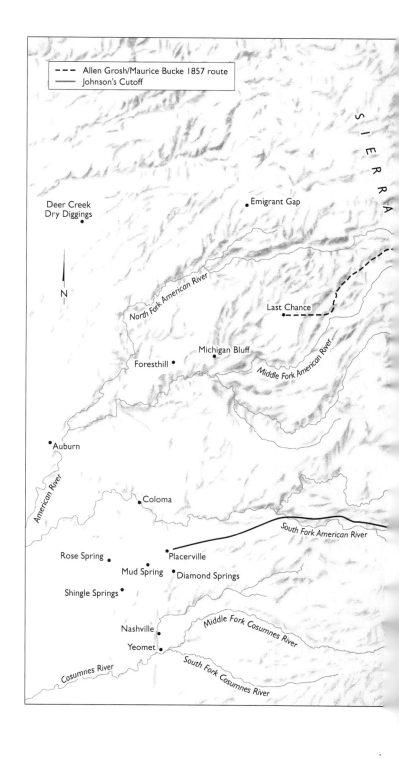

- - - Allen Grosh/Maurice Bucke 1857 route
——— Johnson's Cutoff

Deer Creek
Dry Diggings
•

Emigrant Gap •

Last Chance
•---

N

North Fork American River

Michigan Bluff •

Foresthill •

Middle Fork American River

Auburn
•

American River

Coloma
•

South Fork American River

Rose Spring
•

Placerville
•

Mud Spring •

Diamond Springs
•

Shingle Springs
•

Nashville
•

Middle Fork Cosumnes River

Yeomet
•

Cosumnes River

South Fork Cosumnes River

SIERRA

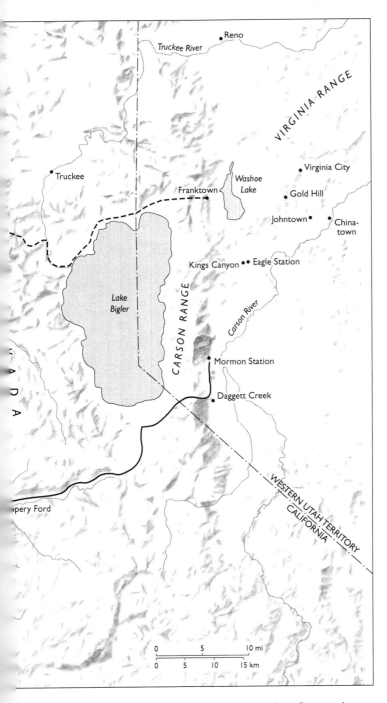

Map of California and Nevada. The Grosh brothers first sought a
rich gold placer in the western Sierra Nevada foothills, mentioning
several locations in their letters. They then expanded their search,
crossing the Sierra east into the Great Basin three times between
1853 and 1857.

INTRODUCTION

"Ho! for the Mountains!" With those exuberant words, the Grosh brothers expressed their innocent dream of a great quest that would take them far from home. Theirs was a nineteenth-century incarnation of a tale as old as humanity: the pursuit of gold in a distant land. As Philadelphia faded from view in the wake of their ship, imagination transformed into reality. For them what was an intimate, personal odyssey would eventually become something they could not anticipate: their letters, which span nearly eight years, captured a changing America and gave voice to tidal forces that would revolutionize a continent.

Ethan Allen and Hosea Ballou Grosh were twenty-four and twenty-two years old, respectively, when the earthshaking news of gold in California reached them in Pennsylvania. The young men were not to be contained. Their lives had been small, and they had yet to travel in any significant way. "The sea!" Allen exclaimed in March 1849. "Oh it is a glorious sight! . . . How much have we lost by not seeing it before." It was only the first of many eye-openers as the sons of the East journeyed to the West. With their trans-Mexico trek, they added a new chapter to the adventure, and although illness tested the limits of their endurance, the horizon continued to expand. Finally arriving in California, they could revel in the phrase that they had "seen the elephant," as people said at the time; that is, they experienced for themselves the great western migration spurred by the Gold Rush during the nineteenth century.

A sense of excitement pervades the early letters, as might be expected. Youth, adventure, and the prospect of wealth fueled an exhilarating optimism that resonated among those who traveled to this new world. Nevertheless, while the wistful images of this chapter in the nation's past are iconic, reality was more complex, and many failed to thrive. Like countless others, the brothers never realized their dream of wealth. Life can be and often is harsh, and Allen and Hosea serve to remind those who would view the past through a romantic lens that a phenomenon like the Gold Rush included many stories. Too often tragedy and disappointment were dominant themes.

WHETHER RELAYING THE THRILL OF TRAVEL and the hope of riches or dealing with the disappointment of persistent bad luck, the brothers were in a position to describe the full gamut of the western experience in the 1850s. Fortunately, Allen and Hosea were educated, and they were devoted to their extended family, which remained in Pennsylvania. They possessed, consequently, the right inspiration and resources to describe their life on the West Coast. Once in California, the brothers wrote extensively about their new home. Besides covering details of their trip, their eighty-two letters include a wealth of observations about mining, politics, the economy, and smaller topics such as how long it took for correspondence to reach them from Pennsylvania. They contracted many ailments and tried numerous remedies, they were attacked by Native Americans, and they considered and pursued various ways to become wealthy. As a result of all this, the Grosh letters record diverse observations not only about important historical events, but also about North American society and culture in the 1850s. Although their primary preoccupation was mining, their letters address a wide variety of topics that could capture the attention of young American men at the time.

In 1853 Allen and Hosea traversed the Sierra Nevada and found that silver—not just the gold that attracted early prospectors—enriched the western Great Basin.[1] First mentioned in a letter in 1855, that discovery earned the brothers a permanent place in the lore of the mining West, keeping their story alive for the next century and a half. The identification of silver helped inspire the preservation of the Grosh letters, but as it turns out, again, the value of the correspondence is more in the form of general observations than in the recordation of specific claims. As a result of three trips into this area, their letters provide some of the earliest firsthand accounts of Gold Canyon and what would become known as the Comstock Mining District. With their deaths, the tragic conclusion of their last sojourn into the Great Basin completes a story that at times reads more like literature than correspondence.

HOWEVER THE GROSH STORY IS FRAMED, Allen and Hosea were, first of all, the product of their home and upbringing. Little in the family background of the brothers would suggest that their story would become wrapped up in one of the great chapters of American history. Their roots were deep in Pennsylvania, and anyone watching them mature likely would have concluded that they would spend their lives within a few miles of their birthplace. Growing up in the Grosh household did, however, provide the foundation for what can be seen within the letters: their family was an ambitious, educated group of people who achieved a great deal. The young men who traveled to California carried the torch, hoping to add a new chapter to the story of success that had defined the previous two generations. And coming from an educated family, they had the tools to describe what they saw and what they thought.

Jacob Grosh, the grandfather of the brothers, was born in a German-speaking colonial Pennsylvania household in 1776. He learned English from his second wife, opening doors beyond his small immigrant community. During the War of 1812, Captain Jacob Grosh led the Marietta Grays, a company of more than one hundred men, to help defend Baltimore. He served for eight years in the Pennsylvania legislature and for nine as a judge for the Lancaster County Court of Common Pleas. Dying in 1860, Jacob had outlived four of his five wives and fathered seven children. His offspring included Aaron Bort, the father of Ethan Allen and Hosea Ballou. The letters mention two other sons of Jacob, Charles C. Pinkney Grosh (which the brothers consistently spelled as Pinckney) and J. A. Bayard Grosh.[2]

Aaron B. Grosh was born on May 22, 1803, in Marietta, Lancaster County, Pennsylvania. He appears in local records as teaching at the Bell Schoolhouse in his hometown by 1822. Aaron later started a newspaper in Marietta with his brother Pinkney and then married Hannah Rinehart in 1823.[3] Aaron eventually accepted ordination as a Universalist minister, serving an important role in establishing his church in Pennsylvania. In the 1850s, he wrote and published what became the authoritative manual for the Independent Order of Odd Fellows, a fraternal organization to which he belonged. The book was a topic of one of the letters, as Aaron's sons reviewed the work and commended their father for his effort.[4]

REVEREND AARON AND HANNAH GROSH named their first child, born November 7, 1824, for Ethan Allen (1738–89), an early Universalist who was also the famous Revolutionary War leader of Vermont's Green Mountain Boys. Their second son, entering the world on August 23, 1826, took a name in honor of the prominent Universalist minister and editor Hosea Ballou (1771–1852). Other children from the marriage followed: Emma Margaret, born about 1829; Mary Letitia, whom the brothers called 'Tish, born July 22, 1830; Malvina French, referred to in the letters as 'Vina, born September 22, 1832; and Warren Reinhart, born about 1834.

Hannah Reinhart Grosh died on November 10, 1849, less than a year after seeing her eldest sons off to the California Gold Rush. Her widowed husband remarried, this time to Sarah S. Smith, on December 21, 1853. The brothers generously referred to the Reverend Grosh's new wife as "mother," a term that can be confusing given that she was their stepmother. Aaron died on March 27, 1884, at the age of eighty, in Towson, Baltimore, Maryland.

A FINE PHOTOGRAPH of Ethan Allen and Hosea Ballou Grosh was taken before the young men left for California, but there is also a record of their appearances in their applications for passports in 1849. On February 24 of that year, they applied at the office of lawyer John Donagan for passports to facilitate travel across Mexico. Reflecting a time before photographs could realistically be required, passports included a description of the traveler: the document recorded Allen Grosh as five foot four and a half inches tall with a medium forehead, brown eyes, medium nose, small mouth, long chin, brown hair, and a dark complexion with a long face. His younger brother was five foot five inches tall with a high forehead, light-brown eyes, prominent nose, small mouth, receding chin, dark-brown hair, and an oval face with fair complexion.[5]

Aaron Grosh characterized his sons in their obituaries in the following way: Allen was "of feeble constitution, [and] confinement to study or in-door labors, for a few weeks, was certain to induce a severe bilious attack, nervous headaches, or deep depression of spirits. Regular school education was impossible."[6] Of Hosea, the father wrote that "in early childhood, he manifested ungovernable passions, which punishments seemed only to increase; and a disposition to exaggerate beyond all bounds, which seemed beyond the power of repression."[7] These were hardly stellar endorsements, but Reverend Grosh was able to concede that as the brothers grew they had assumed characteristics that could be commended.

What might have seemed to be harsh judgment can also be taken as evidence that from the father's point of view, to be a Christian was to strive for ongoing betterment. Reverend Grosh took pride in what his sons had become, and by indicating the weak place from which they started, he was underscoring the great amount that they had achieved through hard work and diligent study. In a final assessment of Hosea as a young man, his father was full of praise: "He was remarkable for his calm prudence, cheerful fortitude, and strict conscientiousness—for his great industry, whether of head or hands, and his steady persistence in whatever he undertook to perform." And in Allen's obituary, the Reverend Grosh was able to add, "but by occasional reading and especially by conversation, he managed to amass a large fund of general knowledge."

Maurice R. Bucke, who eventually became a doctor, prospected with the Grosh brothers during their final year in 1857. Later in life, he wrote a manuscript that dealt in part with his time in Gold Canyon. Although the fate of the document is unknown, Charles Howard Shinn was able to use Bucke's observations when writing his 1896 book, *The Story of the Mine*. Shinn cited Bucke's summation of Allen and Hosea at their prime, remembering them "as of medium height, slight in figure, good-looking, fairly well educated, very quick of observation, ready with expedients, gifted (especially Allen) with exceptional powers of original thought, thoroughly honest and honorable, absolutely devoted to each other, industrious, persevering, chaste, sober, and above all, 'filled with that genuine religion of the heart which is the salt of the earth.' "[8] Raised in a religious household, it followed that Allen and Hosea Grosh felt comfortable quoting from the Bible. On two occasions, they even attempted Arabic to draw on another religious tradition, but they were at the limits of their knowledge and did not manage the Islamic expressions well. That aside, it is possible to imagine the brothers as easy to like and with a great deal of potential.

IT IS WITHIN THIS FAMILY FRAMEWORK that Allen and Hosea Grosh decided to answer the call of the Gold Rush. They linked their fortunes to the Reading California Association, a financial cooperative that combined resources to pay for equipment and transportation to the California goldfields. The stockholders who invested in the endeavor hoped, as one of the brothers wrote, to receive ten thousand dollars in return. The plan of the travelers was to sail to Tampico on the east coast of Mexico, traverse west overland to the north of Mexico City, and then secure passage on a boat for the trip to San Francisco. There was no easy way to California in 1849. Crossing the North American continent was time consuming and risky. A Central American passage required a longer sailing trip and exposure to disease, and going around the Horn was dangerous, lengthy, and even more expensive.

The path the Reading company followed took its toll. Widespread fevers and dysentery left one dead and many sick. This, together with unexpected costs, disheartened the members. Some healthy and impatient participants from Reading left the others to reach the goldfields as quickly as possible. It was a reasonable if ruthless strategy, since the best areas were being claimed by the multitude that was flooding into the Sierra, but it shattered the partnership. When the stragglers finally arrived in California, many were unable to work, requiring constant care as a result of illness. There is no indication as to whether the other members of the Reading California Association compensated the stockholders. As for the Grosh brothers, over the years they managed to send their father fifty dollars and a specimen they hoped might be a valuable opal. (For a discussion of the role of these companies and specifically of the fate of the Reading California Association members, see appendix A.)

Because Hosea was too sick to travel, Allen remained in San Francisco to care for him, diminishing their chance of "making their pile," to use the parlance of the day for striking it rich. The length of time it took to reach the goldfields varied from group to group and from person to person. Certainly, the string of bad luck placed the brothers among the most delayed.[9] In search of at least some way to realize an income, Allen left in late 1849 to help operate a store in Sacramento, but the river flooded and the endeavor failed.

When the brothers finally reached the Sierra Nevada sometime in the summer of 1850, roughly a year and a half after leaving Pennsylvania, they found the region far less exciting than it had been when virgin streams remained to be mined. The fabled Gold Rush had attracted tens of thousands, and the easy pickings were claimed or gone. Placer mining—using various means to wash soil and gravel to reveal the heavier gold—already dominated a broad swath of the Sierra foothills. Still, hard times would not deter the young men: as Allen wrote on January 5, 1851, "Health, youth, strong arms, and stout hearts, and two years experience. What more could we ask? We are in California!" They joined many others who now looked for evidence of underground ore, realizing that although these quartz deposits would require more work and investment, this was the most likely way remaining to gain wealth through mining.

At the same time, the brothers constantly considered and often pursued other options. They tried raising livestock and farming. With hard work, agriculture could provide a living, but it seldom led to remarkable wealth. Finding the occupation a distraction from searching for gold, they returned to placer mining and pursuing quartz veins. Any one of the options they considered might have given them a comfortable, sedentary life, but none of the other possibilities ignited their imaginations like the allure of gold, which remained their primary focus.

Pondering various inventions proved less of a diversion largely because they were always imagining devices that would make mining more profitable. Still, the brothers wrote that their placer digs kept them from developing models of the instruments they hoped to produce. Much of their intellectual energy was given to an improved sluice box for the washing of gold-laden soil. It is unclear whether they ever produced a successful prototype, but whatever the fate of the idea, it failed to yield the profit they had imagined. Another effort led the brothers to attempt the construction of a perpetual-motion machine, something historian Sally Zanjani aptly called "their last and most harebrained scheme of all."[10] Long the dream of the more eccentric inventor, the development of an energy-saving contraption that would require no source of power to remain in everlasting motion was absurd and a failure.

In spite of setbacks as Allen and Hosea moved from one enterprise to the next, they usually maintained an optimistic tone. Naturally, there were times when they wrote of challenges, and they occasionally revealed their despondence over persistent bad luck. On May 22, 1853, for example, the brothers commented that "our continued ill success has sickened us of this country . . . [but] we will not live in the old states poor." They considered their fate constantly but usually indicated that good fortune was right around the corner: "If our luck changes, or rather has changed—as I am persuaded it has . . . ," Allen wrote on May 13, 1850. Invariably, they returned to a familiar theme, that they could only imagine remarkable success, whether it was with one of their bizarre inventions, farming, working a claim in California, or, finally, pursuing a silver mine in the Great Basin.

Besides discussing occupations, the Grosh brothers' letters capture a full array of observations about the West Coast in the 1850s. Allen and Hosea wrote of both a love for and a frustration with California. The brothers delighted in the beauty and benefits of the West Coast, and no matter their depths of despair, these two young men always seemed to bounce back. Part of their positive tone was likely due to the fact that they were constantly attempting to lure their family to the West Coast, and perhaps they simply wanted to avoid admitting defeat. The brothers had settled in a strange new place with remarkable opportunities. These young explorers felt the need to explain this foreign land, while always casting it in the best light so it might attract their family. But mostly, the letters capture how thrilled they were to be experiencing something so exotic and extraordinary. With every turn, it seems, they confronted something that reminded them of the foreign nature of their new home. The fact that Allen and Hosea wanted to make sense of this emerging, vibrant place, to describe it for the uninitiated, is what makes their letters fresh and consequential to this day.[11]

The brothers not only wanted to provide their family with a way to understand California, but also wanted to be part of the process that would create a great state. Allen and Hosea believed in California's potential, and they recognized it was like no other place. The Union admitted California as a unique partner in the nation. It possessed a sudden society cobbled together of many parts, building upon a Mexican foundation, born of the West with its own peculiar perspective and history. The unusual character of this new state was not simply a local issue. This was a time when America was changing. The California Gold Rush opened the West to settlement like never before, shifting the nation's center of gravity. Bound up with this and drawing on deeper roots was the question of slavery. All this meant that California politics was about more than deciding the direction of a new, remote state. The events the brothers described and participated in had national repercussions.

PART OF THE STRANGENESS that the brothers observed in their new home derived from the fact that people were arriving there from all corners of the globe. Because the Gold Rush attracted an international population, the indigenous Mexicans were not the only ethnicity Allen and Hosea encountered. They wrote with some disdain of Chileans, Native Americans, and former convicts from Australia, and they utterly condemned men from Missouri for being uncouth and supportive of slavery. In spite of the range of possibilities, it was for the most part Mexicans who added a foreign flavor to California life, yet they sadly also inspired an ethnic slur from the Grosh brothers: Allen and Hosea, after all, were men of their time. Perhaps to underscore the extraordinary turn their lives had taken, these travelers from Pennsylvania celebrated Spanish importations into their vocabulary, occasionally getting the words and phrases wrong, while employing as many additions as possible. "Dinero," "canyon," "chaparral," and "coyote" are salted throughout the letters. Spelling was occasionally bizarre, and meaning was not always precise, so new were some of these elements to the English language.

Among the travails the Grosh men faced, none were more daunting than health problems. They both had dysentery during the journey across Mexico, and its effects lingered. There were also various injuries and infections during the following eight years. From a historical point of view, the treatments the brothers employed are perhaps of more interest than their actual ailments. They experimented with medicines and ointments, some now known to be highly toxic. Their documentation of uses provides a wealth of information about diseases and cures in the 1850s. For example, Anton Sohn, both a medical doctor and a historian of medicine, suggests that the effort to save Hosea from tetanus was "the first recorded home remedy and medical treatment by a doctor in . . . [what would become] Nevada."[12] As is so often the case, however, letters dealing with areas outside the Great Basin remain to be considered fully.

When it comes to the issue of health, perhaps of even more significance was the way physical challenges revealed the attitudes of the young men. Hosea wrote from San Francisco on March 31, 1850, that Allen had "some ugly stumps pulled from his jaws as it could be done with but little pain, they being loosened by the scurvy." Hosea then added, "rheumatism as usual accompanied it so severe as to draw his leg up slightly though he has it nearly straight again and will soon be entirely well of it. It is well that it broke out here instead [of] waiting till we went to the mines." While the letters frequently captured discouraging turns of fate, they typically included an upbeat note. Teeth had to be pulled, but it was an easy matter because they were already loosened; limbs were bent with rheumatism, but at least it occurred in a city where the condition was more easily managed than in the mountains. Throughout, when they encountered difficult physical issues, they always claimed to have recently obtained the most remarkable solution. Bad news was usually made lighter by a positive spin. Even when faced with their own demise, they told their family that Hosea could certainly be cured of tetanus, and Allen indicated he had found the perfect way to address what was, in fact, a fatal case of frostbite and exposure. Whether it was a desire to diminish the worries of their relatives, a wish to entice them west, or a simple reflection of their cheerful natures, the brothers retained an optimistic tone to the end.

SIMILARLY, ALLEN AND HOSEA launched into the subject of politics with youthful vigor and always with the belief that great change could be achieved to make the nation better. Education had given the brothers the ability to understand broad political issues, and the letters frequently address elections and candidates. At the same time, their religious upbringing and northern roots shaped their point of view. They wrote extensively about attitudes and events in California as the new state took its place in the Electoral College and as it struggled to settle on senators and representatives to send to Congress. And there was the omnipresent issue of slavery and how to deal with the opening of western territories. Because the brothers were determined to be involved in the foundation of the new society, they turned their interests and energies to the emerging Republican Party, seeing it as the best vehicle to accomplish change on the national level and the building of a state that would match their ideals.

The Grosh family's attitudes were wrapped up in their commitment to the Universalist Church of America. The institution had deep philosophical roots in the idea that a loving God would not allow for the creation of souls who would be damned to hell because of the place or circumstance of their birth. In fact, Reverend Hosea Ballou, for whom one of the Grosh brothers was named, went so far as to dispute the existence of hell. The Universalist Church had emerged from the Congregational Church, and many of its members became tied to the abolitionist movement. In 1863 the Universalist Church further established its progressive inclination by being the first national denomination to ordain a woman.[13]

During the critical decades before the Civil War, Reverend Grosh was an avid proponent of the Universalist Church just as it was taking shape. The letters from his sons reveal enthusiastic support for the abolitionists and a willingness to do whatever possible to advance that cause. The brothers spilled a great deal of ink on the subjects of the 1850 Compromise and the dispute that raged over whether Kansas should be a free or slave state. They stood solidly behind John C. Frémont, who ran in 1856 as the first presidential nominee of the Republican Party, founded only two years before to advocate a number of reform-based causes. Their commitment to his candidacy was such that they returned prematurely from prospecting in what would become Nevada so they could garner support for Frémont in California. Whether members of the Grosh family were antislavery because they were Universalists and Republicans, or they were Universalists and Republicans because they were antislavery, cannot be determined. The brothers subscribed to the entire package, profoundly affected by emerging social attitudes, their religious beliefs, and the swirling political fervor of the time. Regardless of what motivated Allen and Hosea to view slavery in a particular way, the subject ultimately proved to be a decisive issue of the American experience during the century, and the observations in their letters provide an opportunity to understand a point of view and the attitude behind it.

Regardless of the conclusions they reached, the brothers tackled contemporary issues with uncompromising enthusiasm, and it often seems that they were far more interested in politics than mining. These were complex, intelligent young men with far-ranging interests. Nevertheless, mining was their vocation, and the idea of realizing profit from the industry was the bedrock of their California experience. One of the more intriguing aspects of the Grosh letters is how they document a transformation that was critical to the evolution of western mining. The California Gold Rush thrived on the idea that small groups of men—and sometimes women—could wash away worthless soil to reveal deposits of gold dust and the occasional nugget. Placer mining was hard work, but it did not require much by way of skill or capital investment. As the '49ers exhausted the best claims, some began to pursue a more expensive, labor-intensive approach. Developing subsurface "quartz" veins, rock that held ribbons of gold, was the next best hope of striking it rich. The process of locating these opportunities, finding the capital to excavate underground, and milling the extracted ore changed the nature of the industry. Because this "hard-rock" mining required training and finances, people usually pursued the approach only when the easier placer claims failed to be profitable. Clearly, a successful quartz operation had the potential to yield far more wealth than washing dirt along a river, but the initial cost and the general challenges of underground mining persuaded most to exploit placer deposits first.

The Grosh brothers were quick to understand the transformation that was occurring in the California gold country. Almost from the start, they sought to locate and develop successful quartz claims. Optimism may have inspired them to overestimate their chances and abilities, but they were clearly looking toward the future of regional mining. In 1853, when they crossed the Sierra Nevada and began prospecting for mineral-bearing quartz in Gold Canyon, Allen and Hosea observed that the Great Basin placer miners were several years behind their California counterparts. The dozens of men who washed sandbars in Gold Canyon were slow to exhaust the resource, while the tens of thousands who hit the western Sierra slope beginning in 1849 more quickly extinguished the best opportunities. Profitable placer deposits in Gold Canyon remained available in 1853, and this was the best chance for unskilled, poorly

financed workers to amass a small fortune. At that time the local miners—
following the pattern established by their California counterparts—had no
interest in chasing after quartz veins while placer deposits remained. The
brothers were, therefore, something of an anomaly when they arrived in the
profitable Great Basin placer district to prospect for rock-bound veins of precious metal.

Allen and Hosea represented a peculiarity in yet another way: they were
searching for silver. Receiving a tip from "Old Frank," a prospector who said
he had found silver in the area of Gold Canyon, Allen and Hosea traversed
the Sierra in 1853 to investigate. Frank Antonio was reputed to have been in
the western Great Basin with a group of Mexicans. Stories describing him as
Portuguese or Brazilian are difficult to evaluate, but what is beyond dispute
is that this prospector returned to California hoping to find someone who
would be interested in his tale of discovering silver ore in the area of Gold
Canyon.[14]

Allen and Hosea were just the sort to act on news from Old Frank and pursue yet another dream. They had a rudimentary knowledge of metallurgy and
a thirst to know more. The books they collected on related subjects probably
placed them at the upper end of literacy for miners in the region. At times
their descriptions of assays and their knowledge of chemistry missed the mark,
but their attempts were admirable. When the brothers crossed the mountains
and began working discretely in American Ravine, an offshoot of Gold Canyon, they identified what they felt was a promising vein of silver ore. The discovery prompted them to spend nearly a year there, causing the longest gap
in the letters, from late 1853 to late 1854, when they returned to their claim in
California's El Dorado Canyon. But they crossed the Sierra twice more, briefly
in the fall of 1856 and then in May 1857, each time attempting to gather as
much information as possible about the silver deposit.

The idea of developing a silver mine may have been almost as harebrained
as their exploration of a perpetual-motion machine. There is an old Mexican adage that "it takes a gold mine to run a silver mine," the reason being
that silver is worth far less than gold, and unlike gold, silver is usually bound
chemically in a way that requires expensive, wasteful processing. Mark Twain,
describing his early days in Nevada, asserted that the proverb was true and
that "a beggar with a silver-mine is a pitiable pauper indeed if he cannot sell
it."[15] The success of the Comstock Mining District, established less than two
years after the deaths of the Grosh brothers, was based largely on the fact that
nearly half the profits of the mines came from gold. Enormous quantities of
silver were welcomed, but gold made the district immediately profitable. The
ore that Allen and Hosea described was predominately silver, and they barely

made a living working local placer gold deposits. The chances were slim that their discovery could have yielded remarkable wealth, given that they probably found only a minor ledge of silver.

Although the brothers failed to profit from what they discovered, they are acknowledged as early discovers of silver in the region. In spite of this, they were hardly the first.[16] Of greater historical importance is the way in which Allen and Hosea anticipated the future of western mining. Unlike most others in Gold Canyon, they prospected for quartz veins. Of course, the need to support underground mining with a large capital investment would have limited their chance of success, but the brothers always strove for achievement on the largest scale, and they were willing to take on this challenge.

With some irony, the newly founded Comstock Mining District became an international icon of underground mining within months of its being established in 1859. The California gold country had been ahead of the curve when it came to prospecting for underground deposits, and indeed, places like the Empire Mine in Grass Valley predate anything similar in the Great Basin. Yet it was the fabulous wealth of the Comstock Lode that motivated prospecting throughout the Intermountain West, not for placer deposits, but rather for riches underground.[17]

Whatever the Grosh brothers identified, it probably was not the actual Comstock Lode (see appendix C). Then they died, taking specific information about their discovery with them, only telling of it in letters to their father and without providing enough information to determine its exact location. In this final chapter, the story of Allen and Hosea functions as a variant of the "Legend of the Lost Mine."[18] There is hardly a region in the West that does not have some sort of similar tale, and the recollection of the Grosh experience provides perhaps the best counterpart of the tradition in the western Great Basin. Indeed, the Grosh story as it is typically related asserts that they actually did discover the Comstock Lode, and this notion persists to the present. Legends of lost mines usually involve a prospector who found an ore body worth a great deal but either died in the process or lost his way and could no longer find the deposit. Many variations occur, and in most cases the mine was never rediscovered, making the situation with the Comstock distinct. Nevertheless, the story of the Grosh brothers draws on the same sort of energy that has fueled this cycle of legends for well over a century in the American West. For those who came to call the Comstock home, beginning as early as 1859, the tale of the brothers seemed like something from the *Arabian Nights*, in which the protagonist asks a genie for wealth but gains nothing because he failed to add that he wanted to enjoy its benefits.

By 1860 miners began to search for what the brothers had found. Someone eventually discovered a hole a few dozen feet deep in American Ravine just south of the newly established Silver City. They named the excavation the "Lost Shaft" and supposed that this was where Allen and Hosea had worked. They also found evidence of assaying in the area.[19] The story continued to resonate in diverse ways. An 1862 map published by H. H. Bancroft shows a "Grosh Hill" just south of Silver City along the American Ravine; it is a remnant of a popular memory that the brothers had passed that way. The following year, the *Sacramento Daily Union* carried a story about the Grosh brothers and the "Lost Shaft." Henry DeGroot mentioned Allen and Hosea in his contributions to the *Mining and Scientific Press* that appeared in 1876. Although this is not a comprehensive list, it is worth noting that the brothers also appeared in published histories by Dan De Quille (1876), Eliot Lord (1883), H. H. Bancroft (1892), and Charles H. Shinn (1896).[20]

Similarly, the Grosh sojourn into the Great Basin earned more than two chapters in George Lyman's intensely researched, thoroughly annotated historical novel, *The Saga of the Comstock Lode,* which first appeared in 1934. And the young men have continued to be a point of interest for Comstock chroniclers to the present. In acknowledgment of their life and struggles, tombstones in Silver City, Nevada, and in Last Chance, California, mark the final resting places of Hosea and Allen, respectively. In addition, a plaque placed at the Last Chance site on September 14, 1985, includes the following: "Here . . . rests one of the two Grosh brothers, the original discoverers of the fabulous Comstock Lode of Virginia City, Nevada." The power of folklore being irresistible, the idea that Allen and Hosea had discovered one of the world's greatest deposits of gold and silver had to be repeated. Whereas historians may hope that the contribution of the Grosh brothers could be properly understood and put in its appropriate context, the popular fascination with their story is untamable.[21]

Of particular note is Sally Zanjani, whose *Devils Will Reign: How Nevada Began* (2006) draws from the letters, providing the most objective and thorough discussion of the Grosh brothers and their role in the silver discovery. Like so many others, Zanjani was most interested in the Great Basin prospecting of the brothers, so their California experience served only as background to her discussion. With Zanjani as a notable exception, most treatments of the Grosh brothers succumb to the romantic haze of the fabled lost mine. And as the focus has remained on the Great Basin, the brothers' observations about the California Gold Rush have been largely unconsidered, leaving the majority of these documents as historical fallow ground.

The legacy of the Grosh brothers included the assertion made by their heirs that they should receive some of the wealth extracted from the Comstock Lode. A weak argument initiated a civil suit, but ultimately, the effort yielded nothing for the plaintiffs. (For a discussion of the case, see appendix C.) The consequences of the legal action aside, it is worth mentioning here that the attempt to gain judicial redress brought the Grosh story to the fore. Without the lawsuit, subsequent historians might not have known of the letters, which might not have even survived.

REFLECTING THE FACT that the Gold Rush of 1849 was an event so monumental, the world looked to Greek mythology and named those who answered the call to California "Argonauts." Jason, the classical hero, assembled the best men of his time to sail aboard the *Argos,* a ship that would take them to a far-off land for gold and for the immortality that can be gained only when a dream is pursued. Allen and Hosea could have lived long but small, forgotten lives, staying close to home with family and friends in Pennsylvania. Instead, like Jason and his Argonauts, the idea of adventure seduced them, and they are remembered for how they reached for riches. Their world was one of dreams, epic adventure, success almost within reach, followed by tragic death.

Separated in birth by less than twenty-two months and following one another to the grave within eleven weeks, these young men ultimately left their correspondence as a legacy, preserved lovingly by their family for one and a half centuries. Ethan Allen and Hosea Ballou Grosh traveled across the Western Hemisphere, survived ordeals, and finally identified an early hint of the wealth that would shake the world just as much as did the California gold strike. The Grosh brothers warrant attention because of all these factors, but it is their ability to use the written word to document their lives and observations of an important chapter of the American story that created their most important treasure.

1849

"Off for California"

Reading, February 27, 1848

Yesterday morning a great crowd of our citizens assembled at the Reading Depot to witness the departure for the land of promise, of as noble looking band of young men as can be found anywhere. They were all in good spirits, and left with the full determination to carry out the intention of the "Reading California Association" to the letter. They will sail from Philadelphia today or tomorrow, on board the schooner *Newton*, Capt. West, for Tampico, from there go to Mazatlan or San Blas on mules, where they will again embark on a vessel and sail to San Francisco. The following is a list of their names: Allen Grosh, Hosea Grosh, Thomas Taylor, Charles Taylor, Andrew Taylor, Uriah Green, Henry Kerper, John Hahs, Samuel Klapp, Simon Seyfert, Peter Rapp, Dr. W. J. Martin, Reuben Axe, William Zerbe, William Thos. Abbott, Robert Farrelly, Noland Witman, and Johnston Flack.[1] The vessel is equipped with everything necessary for their accommodation, and the company with an ample stock of implements, provisions, clothing, &c.

—Reprinted from *Lancaster Intelligencer* (Lancaster, Pennsylvania)

as it appeared in the *Berks County Republican*, March 6, 1849

Based on handwriting it appears this letter is from Allen Grosh:

February 27, 1849

American House, Philadelphia

Dear Father,

Yours enclosing Frank's letter, your correspondence, etc., was duly received and we are very thankful for them. Frank is quite touched with the gold fever . . . and Bayard too! Well, if the country is what it is represented, we will use our utmost endeavors to induce them to emigrate.

Mother has not yet arrived might it [be] possible that the telegraph dispatch would have miscarried? She may come in tonight. If so, we will yet get to see her—which heavens grant. We sail tomorrow morning, and should we not get to see her, I know not what to do.

Aaron Ritter sent word by Mr. Tyson to have our miniatures taken here before we left. But we all were kept so busy that not a moment was to be lost and it was not until this evening that we had leisure to try—we did try however about 4 p.m. today but not succeed. If you can have our images taken off of our other pictures it will do as well.

We sail tomorrow morning, and would have done so this afternoon had not the winds and ice been too strong. We are in excellent spirits—anxious to go. We are very well provided with . . . Indian rubber blankets, tents, caps, bags—good arms, and mining tools. All have worked hard, and the Board of Directors deserve great credit for their activity and kindness.

I have seen most of the friends and they all send love. Brother Thomas is half inclined to the "Gold Fever." He says that were he young he'd go without hesitation.

Tell the girls that the more we think of it the greater is our gratitude for the completeness of our outfit at their hands.

Could we but see mother before we leave we would be satisfied. Heavens grant that we may. We will then be ready to go.

Give love to all.
Affectionately, your sons,
Hosea and Allen

✍ *Letter from Allen Grosh:*

March 1, 1849 12[2]
Schooner *Newton,* Delaware River
Dear Father,

We are on our way to the "land of promise" having left yesterday at 2 p.m. Hosea and I detained the vessel some two hours and would have left us, had not Messrs. Boas, Green, Salliday, Tyson, and I guess all the rest from Reading persisted in not starting until we were on board. Of course we would not have been left behind but we would have been put to the trouble of going down to Trenton. Well, thanks to them all—we'll try and return the favor at the earliest opportunity.

In relation to U. Green I shall do as you request—had in fact made up my mind to try before we left Reading. Have no fear of Hosea and myself on the score of Temperance. There is a small keg of wine bitters on board that is all the intoxicating drink we have and I do not think much harm . . . from it . . . it is so confounded . . . bitter we are very sparing in its use.

Father, I have great confidence in our captain—and the more I see of him the more I like him—he is prudent, calm, and modest—All the men, too, during our stay in Philadelphia and so far on board have behaved themselves very well.

Our officers were selected in Philadelphia—they are as follows:

T. B. Taylor, Captain
Andrew Taylor, Treasurer
E. A. Grosh, Secretary
U. Greene, Director
Dr. Martin, Director

Love to all. Goodbye—
Affectionately and Truly,
E. A. Grosh

U. G. [Uriah Greene] sends love—
I may possibly write again before leaving the capes.[3]

❧ Letter from Allen Grosh:

March 2, 1849
Off the Cape
Dear Father,

"The brave sky bending o'er us
And the wild, wild, sea before us . . ."

Oh it is a glorious sight! Yes, "The sea, the sea! the open sea!"[4] How much have we lost by not seeing it before! Our . . . is leaving everything behind us and we just passed the *Levant* which left for Cape Pleasant . . . the day before we did.[5]

We are all in excellent spirits and bid adieu to land with high hearts and high hopes. Tell the girls that they have fitted us out about as well as anyone in the party.

Farewell. Love to all. Your son,
Allen

❧ Letter from Allen Grosh:

April 9, 1849
Tampico, Mexico
Dear Father,

We at last have everything packed up and will start tomorrow morning early for Mazatlan via San Luis Potosi, thankful that we escape the horror of fleas and mosquitoes, which have nearly eaten us up alive in this otherwise beautiful and delightful little city. Our delay has been owing to the difficulty in procuring horses, Captain West's letter to his agent here never having been received.

The news we receive here from Mazatlan concerning transportation up the Pacific coast, is anything but favorable, and it is highly probable that we will make the journey all the way by land, in which case it would have been about as well to have gone by the way of Cape Horn. However, we are here and if it takes six months to reach the Sacramento [River], it must. We do not want of care on the part of Meyer, Dicke and West.[6] They have done everything that could be asked of them.

Our first three days march will be a hard one—hot and poor water—but after that, we expect a cooler climate and no inconvenience from bad or short water. Our horses are good, and the train of mules is every way in fitting condition. Should we receive information at Durango of difficulty in engaging passage at Mazatlan, I should not be surprised if we struck on to the old military road a little this side of Durango and proceeded direct for the head waters of the San Joaquin. I will probably have an opportunity of writing from that place and if so will let you know.

Our stay in this place, take it all in all, has been a very pleasant one.

Captain West and several members of the expedition, were here during the war and are extensively and favorably knowing; and this has given us advantages over almost any other party that has or will pass this way. The party which came by the *Thomas Walter*[7] from Philadelphia, about a month before we did, were fleeced pretty handsomely, and had not much to thank the Tampicoans for.[8] They were coaxed into betting on the popular game of Monte, and were then fined under a law which <u>forbids foreigners to gamble</u>!—and in a number of ways were annoyed in like manners.

We were fortunate in witnessing the Catholic ceremonials closing Lent, as we call it in the States. Thursday and Friday were observed with much more rigidity . . . in fact on Friday all . . . even the billiard rooms were closed. About 7 pm a procession was formed in the Central plaza, consisting of the military, priests, and all the pretty girls in town, and crowds of citizens, bearing wax tapers, tallow candles, and everything else, combustible, and marched through the town, to the solemn sounds of funeral marches and chants. It was very impressive, and imposing. But it seemed to me, that the fairer portion of the procession were more occupied with thoughts of the living than of the dead—and the Americans were in for a full share of bright glances and sweet smiles, from the sparkling eyes, and pouting lips. On the return of the procession to the church, I was unfortunate in securing a place—but with a little exertion, . . . myself secured a place at the doorway, where if we could not see, we could hear. The chanting by the priests was accompanied by the organ, which was excellently played—and the whole broke up to the many tones of lively and familiar airs performed as a voluntary on the organ. And I must say, that I was better pleased to hear old familiar waltzes and dances, performed as they were, by a willing hand, than all else beside. The city remained wrapped in the gloom of mourning until Saturday at 10 o'clock, when the cannons were fired, . . . and gunpowder burnt, and on every side nothing was to be heard but the cracking of squibs and the hissing of rockets. One, and the most stirring feature of this fast, is the hanging of "Judas" up in the street. On Saturday morning, within sight of our quarters we had some half dozen of these caricatures, hanging about. They are effigies, representing any person you may wish to caricature, hanging by the neck and filled with powder and squibs. At 10 o'clock a.m., on the ringing of the cathedral bells, they are fired, and amid the confusion and noise thus created, another class of "Judases" come into the field. These are effigies mounted on "Jacks", (a small Donkey about the size of 1arge dogs, and . . .). These are let loose and the boys, and the "Judas" and Jack are given over to their tender mercies; and stones, sticks and every sort of missile are hurled at the poor Jack and his load, until the effigy is either torn off, or the animal escapes to the open country. Among the latter class was one taking us off. And to their credit be it said, many of the Mexicans turned aside from it as if ashamed of the affair. The boys generally took it in good patience and laughed at it heartily. But when it came down street toward our quarters one of us pushed on ahead, determined it should not pass our door. They did not come within three squares of us either through shame or lack of courage. Saturday night, however, they received a compliment which fully squared accounts. A party of odd fellows, sauntering through the street about 11 o'clock, having been driven from their beds and all hopes of sleep, were attracted by music to the corner opposite General Urrea's house. The Old General perceiving them then came and personally invited them up. It was a musical party which had assembled there; and there the boys went, with nothing on but their pants, boots, and red shirts. The General introduced them, complimented them, and toasted them. The party sung for them, and our boys sung for the party; and nothing would suit, but that everyone must make a night of it—and a night they did make of it. They kept the whole town awake until broad daylight, and the Old General proved himself to be as big a rowdy as the best of them! General Urrea (well known to our army from his success and skill during the late war) commands this division of the Mexican army in the absence of General la Vega.[9]

Yesterday forenoon I had the first good sleep since we arrived in Tampico, and much I needed it; for I was about worn out. You can have no idea of the swarms of mosquitoes and fleas with which we are infested. It is perfectly impossible to think of sleep at night is almost impossible . . . of having taken advantage of this last 3 or 4 days in writing—for I was too much worn out, and could not content myself a moment with paper and ink, though I spoiled half-a-dozen sheets in attempting it. I am now writing from the top of an old rickety flour barrel, which shakes and jars at every step of those around engaged in packing up for tomorrow's journey. I will not read it over, for fear I may become ashamed to send it, and have not time to write another.

Hosea joins in love to all.

Affectionately your son,

E. A. Grosh.

Letter from Hosea Grosh:

June 13, 1849

Tepic, Mexico

Dear Father,

We have now come to this place which is some 20 leagues[10] from San Blas inland; it is the place where a lot of business of San Blas is done. In coming thus far I flatter myself that I have seen considerable of the Elephant. We will here remain a few days until we can make arrangements for going up . . . coast on sailing vessel or steamer for San Francisco.

We left Tampico on the 10th of April about 10 o'clock warned in advance our journey through a country reported to be very barren in water some days we would be obliged to encamp without water, this last proved groundless for we always had water though sometimes very bad. Our first evening was a sudden introduction to the life that we were to lead while on the journey. It was late in the afternoon when we arrived at our stopping place, Altamira, and by the time cooking was commenced, it was dark; All was confusion and botheration, we cooked our jerked beef as well as we were able, and ate our ship biscuit with it and our coffee as well as we were able in the dark without any light but our fire; the chief of our mess was sick . . . The next morning we were roused and ate jerked beef with ship biscuit and coffee. The main body got off about 6 o'clock, the rear guard of which Allen and myself formed part a little after 7. So now we are under way. I will not fill my letter journal fashion for I would fill my sheet before I was half through if I should put all I have noted on the road and make my letter less interesting than to do otherwise. I will now stop for tonight, as it is late and I am tired with looking over the treasury accounts, (for I was elected Treasurer at San Luis Potosi) and finish in the morning.

14th. To follow the mules all of one day is a lesson of patience that is not to be met with elsewhere one of the mules lies down another slips his load another takes his course through the chaparral the muleteer lifts him up reloads or drives him . . . without scarcely a murmur or any exhalation of breath. During the first few days we had several stampedes. Fortunately no one was severely hurt. It was always an exciting time when McDowel, a splendid horseman and all the Mexicans started off in full chase, always capturing the runaway. A full description of one day will give you some idea of the rest until we came to San Luis Potosi. We started early sometimes as early as 3 o'clock after a little cold breakfast if anything remained from supper if not with a cup of coffee alone, to avoid as much as possible the heat of the day, generally arrive about 10 or 11 o'clock, then to look for a place to cook while sheltered from the sun, then the cooking, then to lunching as the ship biscuits lasted we had bread but it was gone before we had been out a great while and then we had to boil mush which added much to our labor.

Tuesday April 17. There being a river to ford we did not intend to start till daylight but were delayed by the loss of some horses until past 9 o'clock. We left West with the sick while we went on. Just before we came to the river we came to a fine grove of trees which were the first we had seen since we came to Mexico that gave a good shade, shortly we crossed a river, the Tamesí[11] and stopped at a place called Le Monde and waited for further orders from rear guard and baggage. News shortly after arrived that the baggage had by another road passed us. Orders were immediately given that we saddle up. We started off in the hot sun and rode about four miles when I discovered that I had forgotten my ammunition bag and powder flask, contrary to the advice of all I rode back at full speed and fortunately found them where I had left them. Now I had nothing to do but catch up to the party which was no small job. I started as fast as my horse could carry me and after getting once or twice on the wrong road for a short distance and cutting across to the right one and galloping and trotting for about 8 miles I succeeded in coming up with one of the men whose horse had given out and one of our Mexicans when I felt relieved as I was certain to keep the right road and catch up to the party however late. We came to the place where we expected to find the party, . . . part way up the mountain the horses led by the Mexican and our man refused to go so I went on to come up to the party and send them back help. When at the top of the mountain I found two roads and took the one to the right after following it for about 3 miles the road became very indistinct and finally divided into cow paths, I then tied my horse and walked back when I came to the edge of the mountain I found from some inquiries of a couple of Mexicans I met that I was on the wrong road, and went back to my horse and came back to the road where I started, just at dusk. Then I had to go down the other side in the dark. Trusting to my horse to find the path in the dark I pushed on at a slow walk until nearly down I found myself completely off the road. There was nothing to do but to stay where I was or run the risk of breaking my horse's legs over the rocks, so I unsaddled my horse tied him by the lasso to the stirrup so that any start of his would wake me. I lay on it for a pillow my carbine leaning against a rock ready to my hand and composed myself to sleep. I had scarcely got to sleep when I was wakened by the barking of dogs about ½ a mile off. I immediately arose and started on foot for the place after taking a glance at what stars were in sight so as not to lose myself. I reached it after 15 minutes rapid walk through the chaparral, I found our fellows encamped. They had been out for me but meeting those who had been with me were told that I had gone back where the mules were so felt perfectly easy about it. We had left Tampico but a few days when Allen came down with diarrha[12] (I declare I can't spell that) which weakened him very much. The day following the one that I was lost he was so weak that after going 5 miles he gave out and was obliged to stop. I stopped with him and took him by ½ mile stages the remaining 4 miles after intervals of rest. I thought at one time that he would not get through. We arrived about 5 o'clock. He cured his diahrhea some time after but he's not been well since. . . . he started being troubled first with weakness then fever and ague[13] and weakness. He is now much better. A few days on the ocean will make him right. As for myself my health has been good except that at San Luis I was troubled with sour stomach and weak digestion the prevalent disease of that portion of the country. We have had a good deal of sickness among us nearly ½ of us being at one time or other unfit for duty. At Buena Vista (one of ½ dozen places of same name known only by the route which they are on) we lost one of [our] men by congestive fever, had another carried for several days on a litter, and then 3 of them in a carriage for one day after which we lay at San Luis Potosi where when we left, we left Abbott behind who did not overtake us till four days after we left Guadalajara. The man we lost was our Treasurer A. Taylor, the captain's brother.

Until we reached San Luis we had little else to eat but mush and pork fat,[14] at San Luis we had a little bread. Had a good deal of genuine concern about doing our own cooking which is a severe task no mistake. . . . At San Luis we bought a couple of wagons which while it supplied a want occasioned new trouble.

In traveling through this country one is struck by the scenery. The sharp and bizarre outlines of the mountains though little foliage is seen even on them until you approach toward the Pacific when the timber becomes very fine. A great proportion of the eastern part of the country is barren even such trees as there are afford but little shade. We generally did not find water from starting to stopping but few trees were seen some days none except the bushes called chaparral and the different varieties of cactus and prickly pear which 15 or 20 feet high forming in many places the only trees. Some days we went over level plains covered with dried grass when our eyes were refreshed by the mirage that gave appearance of water spoken of by travelers over our . . . and eastern deserts.[15] As this is to go this afternoon by private hands I am hurried and therefore can write no more at present, but will do so more fully bye and bye if I can send it by vessel or steamer on the Pacific.

Please get the directions that Dr. Behne[16] promised to give us but which in our hurry we forgot to get and send them on to us.

Your affectionate son,
Hosea

Though Allen is not sick still he feels hardly able to write especially as he has a letter to write to our directors in his capacity of Secretary. In fact none of us feel as we did at home even though perfectly well, indeed it is no easy matter to get pen ink and paper or a place to write. The latter particular accounts for this being so scratched and scribbled.

Extracts from a letter written by Reading California Association secretary E. A. Grosh to the president and directors of the local company:[17]

June 29, 1849
San Blas, Mexico

Captain West failed to carry us through and we are now left on our own resources, and will have to go up the coast without aid from him. We came from Tampico, 50 miles, by our own means. The extravagances of the Captain's partner and the insufficiency of the $200 passage money are given as excuses. Some of our party are sick. At San Louis Potosi we had to get conveyances for the ill, as some are unable to ride. Hiring a team would be $175 for 90 leagues. Buying a coach would cost from $1,500 to $2,000. Then we bought two wagons and eight mules used by the army during the war. At Tequila, two days from Guadalajara, we were overtaken by Capt. West, Farrelly, and what was most gratifying of all, Uriah Green with Abbott in charge. We had never expected to see Green again. After almost insurmountable difficulties we reached Tepic with our wagons. Here the expedition failed. Capt. West said he did not have any funds for another day's provisions. From then on we were on our 'own hook.' All but the sick walked. The last day we were above our knees in water. We are here at an expense of $5 per day each. If the fare is not more than $50 per man, we will take passage on a ship that leaves on July 24. If it is more, we must get there by some other means. We will take our wagons and mules with us if we can. There is a good market for mules. Our sick are Dr. Martin, Charles Taylor and Abbott.

℘ *Letter from Hosea Grosh:*

August 31, 1849
San Francisco
Dear Father,

We arrived here yesterday morning with a light southerly wind. I wrote at Tepic but still as that letter may not have reached you I will say that after some delay we arranged all our affairs at San Luis Potosi bought wagons there and started on leaving. Abbott then very low but in good hands with the best medical attendance, U. Green remained with him. At Guadalajara we waited for West and Farrelly who had gone back to try to recover some stolen horses at the end of some days we pushed on again about 20 leagues from there we lay by for our baggage to overtake us (we hired mules to carry our baggage as the roads were impracticable for loaded wagons; in fact we were told that we would have to take our wagons apart and have them carried over at a place called the barrancas)[18] at a place called Tequila. Who should come in but Abbott and Green, West and Farrelly and the baggage guard all in a heap nothing of any very great importance occurred between there and Tepic except that a company[19] from Connecticut overtook us and went into Tepic with us on the 11th of June. We remained at Tepic until the 20th when, as we expected the steamer on the 24th, we started for San Blas. The road was abominable, almost impassable for wagons, one day from daylight to dark we made but 3 miles. We reached San Blas on Saturday [June] 23 in the evening. We managed to get up the steep hill on which the upper town is situated and got a place to stay at the usual rate, 25 cents per room. We found the sand flies an awful pest. Their bite seemed more like a burn than anything else. At first we thought the mosquitoes bite the worst but we got some used to the bite of the latter while the former grew worse and worse. . . . so we were obliged to wait a chance for a place in a sailing vessel. Six of our men asked for their share of the cash in hand to pay part of their passage in the brig *Venado* then lying in harbor and up for San Francisco. We advanced their share of the funds and they paid their passage. We advanced the same for Farrelly and Seyfert they making up the balance from their own funds to pay their passage in the steamer. We were anxious to have them go as we could depend on their ability as well as honesty and thought that perhaps our goods might have reached there. The night before the *Venado* was to sail she was struck by lightning and one man (the interpreter of a party from Mobile) killed and one severely hurt. Our six were uninjured but soaked to the skin by the drenching rain. They came back to us next morning without money. We . . . took them in but our funds were lower than ever by near $350. . . . at that time I could not have written to save my life yet I did <u>not</u> have the blues or lose my hopes of getting out of all our troubles and but for our sick we have been light-hearted. On the 9th of July the *Olga* came in from Mazatlan expressly for passengers for San Francisco. We engaged our passage in her as steerage passengers the captain to find us and paid for six passages and gave our wagon's harness and arms as security for five more.[20] There were but eleven of us (one of the shipwrecked having engaged to work his passage on board the bark *Hortensia* to sail on the 1st of August as carpenter). Captain Taylor with Kerper and Klapp had gone to Mazatlan with the mules hoping to sell them to better advantage than at San Blas and the two that went in the steamer just filling our complement. We were now rid of a place where no work was to be had; in ordinary times the abode of about 300 "greasers" and myriads of sand flies and mosquitoes. One thing however we did have the most splendid sea bathing . . . at Tepic. West declared all his funds gone and we must take care of ourselves which announced the expedition done a complete failure. On the 12th of July we sailed for Mazatlan where the Captain expected to take on board . . . other passengers. We arrived there on the 14th and left on the 16th. In the course of those three days I found or at least thought so by his conversation with some of his friends, that we had a Universalist preacher aboard. I asked him if that was the case and found him: the Rev. Mr. Bull of Indiana.[21] (who I see in your letter is mentioned as one of the preachers who had gone to California) I introduced myself to him and also Allen. He appeared glad to meet as we certainly were to find him. It seemed like finding an old acquaintance. The following Sunday we had preaching it was right pleasant to listen once more to preaching especially Universalist preaching. We had preaching every Sunday until the one on which we ran into Monterey. After being out a short time from Mazatlan, I was taken with chills and fever but I let them take their course without troubling the Doctor but dieted very closely and got rid of them. For a time got rid of them and again got them about a week ago and have had them to the present time. The cause seems to be the coldness of the weather even a short distance from land in these seas; on the whole I never passed so cold an August. It was more like March. This with the effect produced by the low land and marshes of San Blas probably was the cause as I escaped all diseases except sour stomach coming through Mexico. And while there, Allen's health however has been good and I think mine will be in a few days. We lay a couple days at Monterey, which we were 5 days in getting in, after making the land 40 miles below, on account of the fog, while there the weather was quite warm, the day before our arrival was very warm while we within ½ a dozen miles were suffering from cold, I had no chill while there and felt better . . . a fair wind (the only one during the whole voyage) which though light brought us up in little more than 36 hours. We have now been here two days and they are not here yet and most of them will not be before day after tomorrow. We received your letters, at least I did, this morning and it was a real treat to hear from home. I have not had time to write before this evening. Immediately after our arrival here the vessel containing Taylor, Klapp, and Kerper came in. Two of our men stopped at Monterey to work at $1 per hour for a few days intending then to come up by the steamer to meet us. We met Farrelly and Seyfert here. A few words more in relation to San Francisco and I must close. I will write more at another time. It is tolerable pleasant in the morning until 9 or 10 when the wind begins to blow and it gets very cold. Work is plenty and good so that though we have little money we can get along in spite of high prices very easily. The place is growing like a mushroom, houses springing up some with clapboard sides and canvas roofs others all canvas and long lines of tents in every direction a large proportion of stores, grog shops, gambling establishments (which two latter are very plenty) are in tents. Gambling is carried on extensively and on a large scale. Wages vary from $3 to $16 per day; board from $2 to $3 per day. I must close for the present but will write more at length by next steamer. Mr. Seyfert is afflicted with diahrea and is going to return in this steamer. By him I send this. Allen is so busy and writing so inconvenient that I do not think he will be able to add anything. Please ask Dr. Behne what those directions he tried to give us to avoid bilious diseases as we were so hurried that it was neglected. I have not referred to your letters or newspapers particularly but I have read them both with the greatest pleasure. Thought of Mario . . . also. Remember to all.

As ever your affectionate son,

Hosea B. Grosh

I would just remark that nobody steals anything here. As to the mines the accounts are contradictory showing it to be something of a lottery in wages. Carpenters rank highest $12 to $16 per day. No business is considered to pay unless it yields $1200 per month. Allen was at the printing office yesterday though as to the result of the call I have not seen him since to speak to him. Dr. Martin of our company is quite low with dysentery. Flack also has it badly. Charles Taylor is weak yet but not sick. Abbott has the chills within a couple of days, but escaped yesterday, otherwise pretty well. The rest are in good health at present. Should we see Dr. Osborne we will attend to your money matter. No doubt he will. All men are liberal here and he is not miserly you know. On your birthday I remarked it to Allen but when mine came I did not think of it till past. We will go to the mines as soon as we can make provision for the sick and raise the money. The fare is about $15 to the highest settlement each passenger is allowed 100 pounds of baggage. I have put in these things in item form as there is not time to digest them. I was afraid I might forget some of them. H. B. G.

ᴄᴏ *[The following is lightly written and then crossed out.]*

Dear Father,

Please wait for Ms. Sallade and Mr. Taylor and give them the news as I am too busy with the disembarcation of our goods to write. My health has improved wonderfully. I can say no more than love to all.

Allen

ᴄᴏ *Letter from Allen Grosh:*

September 29, 1849
San Francisco, Alta California
Dear, dear Father,

I will be compelled to cut a long letter short, for want of time, not for material. Enclosed you will find a letter from Hosea, written before we left the Barque *Olga* of Boston in which vessel we came in. The reason I did not commence earlier is that I do not like to write bad news, and though I do not give you good, yet it is better than I would have been a few days ago.

Hosea has been very sick, though improving and in the course of a few days, if no accident happens will be able to be about again. Dr. Walter J. Martin of Allentown died on the morning of the 13th inst.[22] of dysentery; he had been sick almost from the time we left Tampico and at last closed his eyes in that land he so anxiously wished to see. This disease the physicians assure me, is, in its last stages, very contagious, and, as you will see by his letter, Hosea was suffering from the fever and chills when we landed. The whole care of Martin devolved on Hosea and myself and immediately after his death, Hosea's chill changed to dysentery and in a day or two he was on his back, despite injections, "No. 6"[23] etc. I never heard of anything like it—it was so sudden, and for a while I was very uneasy. What made the matter worse—his confounded patient, uncomplaining disposition, prevented me from discovering how bad he really was. Judge, then, my consternation, when, on his complaining of being tormented with visions on closing his eyes. I felt his pulse I found it indicated violent fever! His physician, Dr. Henry B. May,[24] of Boston, (an Eclectic) was also suffering from the same disease, and had not left his office for several days merely prescribing from the symptoms as I gave them. I told him, now, my fears and insisted on his coming over, it being but a few steps. He did so, and was as much frightened as I had been. . . .

Hosea of course is much reduced, not being able to sit up for any length of time—But I have no fear as to his recovery, and I beg you and mother do not make yourselves uneasy about him. A few days ago, and I <u>was</u> fearful—I did not dare think of it—and were it any other than Hosea I would be fearful still, for home sickness in his present state, with its serene song of past joys and memories, would shake a less patient and cheerful soul terribly. Ah! "Home, home! sweet home!" Father, has a strange, a wild, fearfully deep meaning, out here, which it is hard to understand while sheltered by a paternal roof! And I have seen more than one stout heart wither, and more than one strong arm fall palsied by thoughts of home. Dr. Martin fell a victim to this insinuating, fascinating, mysterious disease, whose blow falls and you know not from whence it comes, and whose demons are bright and happy faces. (See next sheet)

(2nd sheet)

(I destroy one half the other sheet, not having time to finish what I began.)

I have been taught much wisdom and but little happiness by this long march toward the setting sun, and things that once appeared bright and glittering now show themselves but a black unshapely mass. It is indeed cruel to teach enthusiasm reality—truth—it felt so cold so dead upon the heart as well as to crush it. Oh! how many bright dreams of youth have faded away before the stern reality . . . to my dismay is the discovery that I have been deceived in my companions and more than once have your warnings before I connected myself with this company raised up before me. I have been greatly deceived, and I verily believe that before next spring Hosea and I will be the only ones of the whole party revering in California—or at least faithful to the stockholders. I would give almost anything to be free from my obligations, so would Hosea, and I scarcely know what to advise our friends among the stockholders—everything is confusion—the company as good as broken up. It depends a great deal on fortune how we succeed—if not unfortunate in the mines this fall and winter, the stockholders will get all they risked and something more. If not, it is a bleak look out. Hosea and I have made up our minds come what may, we will do the best we can, and for two years we are theirs—after that time we are <u>free</u>. I can do no more, neither can Hosea. As long as Taylor, Green, Farrelly, Zerbe,[25] Kerper, and probably two or three more, remain here, they will be true, too, but they are all heartily sick of California.

Our party is scattered very much. Hahs and Witman stopped at Monterey on our way up to work—the last, probably, that will be heard of them. Kerper and Klapp went up to the mines a couple of weeks ago—this morning Kerper came back, sick, not having been able to work at all—Klapp worked hard for a whole week and earned $7 gold. He is here. . . . Rapp started for the southern mines a day or two ago, and perhaps will compromise with the stockholders at a future day, if successful, for I do not think he will have anything to do with Taylor—neither will Klapp. Last Tuesday Taylor, Green, Zerbe and Abbott started for the San Joaquin mines. Taylor and Green may probably work together but Abbott and Zerbe will fall out with them at "the first opportunity." Both were dissatisfied with Captain Taylor—and I do not think Abbott over honest. Charles Taylor returns home today. Flack and Hosea are yet on the sick list—Flack nearly recovered—and remain here with Farrelly and I. Axe we left at San Blas to come up in the American Barque *Hortensia*,[26] which has not yet arrived—Seyfert returned home last steamer.

I have been very much disappointed in Captain Taylor, and if anyone is to be blamed with the breaking up of the party, it is he. I was sick coming across the country, Hosea and Farrelly were new members of the Board, and Green and him done all the business, and made many mistakes as to lose entirely the confidence of the men. Add to this Taylor's hard and overbearing disposition and you will at once perceive the result. The men were not treated as they should be.

Before they left, Taylor managed to kick up a row. . . . We managed to raise $1000 . . . horses for $1200, and when they left there was less than $800 in the treasury, and some outstanding debts reduced the amount to about $650— They wished to take upwards of $400 with a month's supply of provisions, along. We objected and asked them to pay at least $60 of the debt. This was the night before they started. Well, so matters stood, until next morning when some slight misunderstanding about a blanket occurred between us, and before I had time to think of it he [Captain Taylor came] down upon me— giving the lie, etc., then leaving the subject of dispute [that is, the blanket], accused me of aspiring to be leader of the party! (a station I would not occupy for $10,000 a year)—and informed me that I had been an encumbrance to him—had tried to kick up a row with him, ever since we had been here and a great many other things too numerous to mention and equally true—ending with a polite intimation that he did not want me to go up to them, and a threat to strike me—which I, of course, dared him to do—but he didn't, though. However he slipped off without paying his share of the debt, and managed to give Farrelly the hint that he did not want him in his party either. Farrelly, Hosea, and myself are the proscribed ones—because we did not approve and sustain their (Green's and Taylor's) foolish way of doing business. So the matter stands. Hosea and Farrelly both urged me to lay the whole matter before the stockholders, together with the manner in which he left me here with the sick in charge, so short of funds. But I have thought otherwise—and will write him a cool calm letter, reviewing the whole ground of dispute, and if he refuses the olive branch lay the letters with his answer before them. . . . But of course rest assured, I feel perfectly easy as to the consequences, and in every important point in the dispute I have strong proof against him and sufficient to clear myself of all blame. In fact, the only thing I am to blame for is too closely adhering to and sustaining him in measures I did not approve of, simply for the sake of discipline. By next steamer I will write at length on this and other subjects. We have had so much sickness since our arrival that I am compelled to forgo the pleasure of writing to my friends at home and cannot even notice your letters as I should wish to.

Living is very high—and though we are short of funds I feel in no way uneasy—as Farrelly has a good situation at $12 per day—and will divide to the last penny—Farrelly, Seyfert, Hosea and I were fine friends throughout the journey—and though we have lost Seyfert "we three" stick together like wax.

My health improves every day—and I am one of the few who do not regret their coming to California. Hosea and I may not make anything the two first years—but after that we will.

Your son, Allen

❧ *Letter from Hosea Grosh:*

October 28, 1849
San Francisco
Dear Father,

In my last to you from on board the *Olga* while we were lying in this harbor my excuse for writing such an unconnected letter was that I had the chills and my condition in the present instance is far from being improved in comparison with that for upwards of six weeks. I have had dysentery which of course has in no wise impaired my powers of mind which have so close connection with the health of the body. On the Sunday following . . . the death of Dr. Martin who . . . contracted dysentery, of which he died on the 13th and we buried him the following day. Being now free from the necessity of continual exposure to cold and change, which the attendance of Allen and myself upon him had involved, I now thought that the best thing I could do would be to do something for the ague which had clung to us this time I accordingly commenced using a remedy which the doctor assured me had not failed in a single instance since he had been here. The evening of the 14th of September the dysentery commenced on me in a severe form. I used injections which Allen and myself thought calculated to check but with little or no effect. On Sunday morning I ceased using the ague medicine, which I think increased it, the more readily as that disease seemed to be worsening and called on the doctor. For a considerable time there was but little change that on the whole could be considered an improvement I knew by the doctor's manner that he considered me as a doubtful case at least, perhaps worse, but during the whole time my spirits were very good, so much so that Dr. May every morning when he came in thought that I had improved until he felt my pulse. He inquired my symptoms when I noticed his countenance was lower than ever, if possible. At last I began to improve and all went along quite well almost ever since though with one or two little backsets arising from eating what did not agree with me. The weather, which changes much, not from day to day, but from day to night, and other problems . . . so I thought that though not fully recovered I would send a few lines as you would be better able to judge of my health by this than by any description that Allen might give you. I am during the past couple of days been improving more rapidly again which I attribute as much as anything else to a fine spell of weather which promises continuance for some time. I am able to be up and move about, walk to see the doctor and wherever else I feel inclined though I am still quite weak. Allen feels so much encouraged about me that after tomorrow he intends to look out for a situation of which there are plenty though the pay may not be more than $100 or $150 per month which is considered low here, and but few stay longer than just to procure enough to carry them to the mines unless their health prevents them for so doing. I am in hopes to be able to do something soon also.

Oct 30. I would have been content with what I have already written and sent home a ½ sheet as my share of letter, had it not been that since I wrote, there has been a marked improvement in my health and strength, in fact I find myself as strong or nearly so as before I was taken with dysentery, though I had not at that time my full strength, as I was weakened by fever and ague. Every other day I am better, but it seems that on my well day I have other things than letter writing to attend to, so that I did my writing on my worst days, day before yesterday and today. Yesterday I noticed that I felt far better than day before, today I feel fully as well, and much stronger than even on yesterday and this too in spite of a cold that I have caught, how I know not. Allen has told you of the difficulty that exists between himself and Captain Taylor. . . . we are now . . . to settle according to our contract with the stock-holders at the end of two years from the 1st of this month. After that what we make will of course be our own: this course is the only one that is consistent with honor and honesty, that gives any hope of anything like pleasure, or comfort in California; though I should not be in the least surprised if in one year from this if there were not a single member of the company in California except Allen and myself in fact from present appearances the contrary would surprise me; as for us I think that in the course of four or five years at farthest if not troubled with much sickness we will be able to collect together some $15 or $20,000 each with which I can feel well contented to return home if I can get more in that of course I will do so. Though we calculate to remain here until we do make fortunes it is altogether probable if not certain that we will be home on visits more than once before the time is up.

Those who suffered from sickness through Mexico are almost invariably hearty and now generally in better health, strength and flesh than ever they were at home, some of them so fat that their friends at home would scarcely know them; this gives reason to hope that after this spell is over my health will be firmly established, and being acclimated, that there will be little further danger for Allen or myself. Samuel Klapp came from the mines a couple of weeks ago bringing news that Captain Taylor lay at Stockton sick with dysentery, for a week, from which he had just sufficiently recovered, when he came down, to continue the way to the mines with Green, Abbott, and Zerbe. Our latest dates . . . home are in latter part of June the mail for September, October, and I think August have not reached us so that you may well believe that we look anxiously for the mail which we are informed is on the way (the Steamer carrying having been delayed by getting out of fuel so that the *Senator* which six days later started from Panama arrived here four or five days ago and brought news of her so that we will hear from home soon unless our expectations are sadly disappointed. Our company, as Allen has doubtless told you, is virtually broken up as far as actions in California is concerned though I do not . . . think that the stockholders will lose anything eventually though their expectations will be disappointed as regards the profits of the "Spree" which were exorbitant even considering the exciting accounts daily received at that time and published from one end of the nation to the other and appear still more so when we compare them with others . . . Of those who came with us a few have returned a profit each day while most of those heard from have not much more than paid expenses, some, not even that; Though all agree that gold digging and washing must be learned like everything else must be learned so that the future may enable me to give better accounts. If I feel like it and have anything more to say I will add a P.S. tomorrow if not.

Good bye from your affectionate son,
Hosea B. Grosh

P.S. Will you ask Dr. Behne for the advice he promised us but which we in our hurry forgot to call for I am not certain whether I mentioned this in my other letter. By next mail I will be able to send the stockholders a full report which will be accompanied by my resignation though they have never confirmed my election at least I have received no intimation of it as yet.

Letter from Allen Grosh:

October 31, 1849
San Francisco
Dear Father,

Hosea speaks for himself this time so your anxieties about him will be removed.

I have been waiting in vain for the arrival of a steamer for this past two weeks (3 are due from the States) and all California is excited to an extraordinary pitch by the gross negligence of that monster monopoly the Pacific

Mail Steamship Co. It is outrageous! And if government will do nothing in the matter, California will do it herself. I shall therefore devote this letter to the future. . . . I should like above all things to winter in the mines and would do so yet if I was not fearful of Hosea's health.

For this last three months I have been planning a machine to wash gold, which I am certain will wash almost as clean as quicksilver, and about 5 or 6 times as much earth in a given time as the common cradle.[27] It passes through two entirely different processes. When I get it fully planned I will send you drafts. I will be able to work it with quicksilver should that prove of any advantage—and one pound will go as far as 5 in the common machines. One thing I would mention as curious in connection with this subject—I met Dr. Evans of Pottstown and in conversation soon discovered that he had struck on the same chain of reasoning, and as far as he went, followed closely in my footsteps—but he stops short of the mark and will not succeed unless he profits by my hints. The principle is this: the specific gravity of gold is about 17, and that of the heaviest substances found along with it is not 9.[28] The application of the principle is to use water as a medium through which to pass the earth by its own specific gravity thus allowing the heavy particle to reach the bottom first then closing off the lighter ones. His plan is to make a pass with water of the earth to be washed, and in that state pass the remainder through his machine. There are two problems with this. First—The quantity of dirt worked over by this process will be too small to make it profitable. Second. The clay or adhesive quality of the earth being present—a fall according to specific gravity is doubtful.

My plan is to use a trough (or series of them for more extensive application) based partially on those used in the tin mines of Cornwall (I made a draft of one at home before starting, having got it from Mr. Nichols)[29] by which I would wash away all the lighter particles, leaving only the black sand, gravel and gold, which is then subjected to the second operation, described as Dr. Evans'. The power used could be either water, steam, or hand, and extensive application for the employment of 10 or 12 hands would of course preclude hand labor. A single machine of 4 troughs 9 feet long and 4 wide would afford employment for 2 mules and carts and our whole company. What a pity that things should so turn out with our company. However, it is not my fault and as Captain Taylor has been pleased to charge me with a desire to supplant him in his command, I have declined the services of any one belonging to the company unless it be Farrelly who is also included within the scope of his displeasure, having been very politely informed that the company was large enough without him. The whole truth of this matter is, that I was thought more of, generally, by the members than he was, and was ill less often than him, . . . The machine is complicated, and will cost something, but I will build it myself. I will get a model under way this week yet.

My present plan is to get a situation at anything I can do, and one that I can if possible give up at short notice, so that I will not be embarrassed by any engagements should I have an opportunity to go up to the mines—Hosea's health permitting. No doubt I can get from $100 to $150 per month, and that is better than doing nothing though it is not making a fortune. The rainy season will not set in for a month yet, probably not for two. If I go to the mines I will go up the Sacramento [River].

If I had had five hundred dollars of my own when I came here I would now be worth probably a couple of thousand. There is every chance for making a fortune here, but you want a start. We will get that by and bye. I had a great mind to write for John Jones, when we first arrived—but I am glad I did not—the acclimation process is pretty severe. Besides his business of mason is not so very good, all the houses being made of wood. But baking is excellent. Flour is not expensive. . . .

Brick is the cheapest building material we have ($40 at $45 per thousand) but for want of lime it is not much used, cement from New Jersey and Great Britain is the substitute. If I had five hundred dollars and could get someone in with me who understood the business, I would go to burning lime at once.[30] There is good limestone in the country, and you could sell it at almost any price you chose to ask, say $3 per bushel. But wood will not answer for building purposes here. Stone or Brick must be used. About half of each day it blows a perfect tornado from the northwest—some day or other a fire will break out, and in half an hour San Francisco will be a heap of ashes. A single glance at the place would convince you of it. This will probably happen before next spring. And if John will come out by that time I think he will do well. I know he will do well. The brick business, if he understood that would not be a bad speculation. But if he does come, let him bring every cent he can raise. Money is worth at least 10 percent per month here, and cannot even be got at that. If he wishes to escape sickness he should come soon, for it is much healthier now than it will be in summer. Once acclimated, I am assured by old residents, with a little care as to diet, there is no danger. My own health is better now by far than it was when we left Reading last spring—and I have endured trials that would have killed me at home. Yet, many do not like it here—for every day I see those going home, dissatisfied, who landed here one month ago without a cent in their pockets, and who in that short time have made enough to take them home by the steamer. San Francisco and the mines are the most disagreeable parts of California. Monterey, only 60 or 80 miles down the coast, has as fine a climate and is as healthy as could be wished. Two of our members stopped there and stayed about a month. One of them is now here, and he said he would as leave live there as at any other place he ever was in. San Jose is equally delightful from what I can hear. So is San Diego—Los Angeles is the finest place in the State, as we insist in calling her.

If John chooses to come out, he cannot fail in any branch of his business. I would like to urge him—but dare not. But this much I say—If I make anything by that time, or can do anything for him, he can command me to the full extent of my abilities. If at any time he should make up his mind to come give me notice and I possibly may be able to have all things ready for him. Judging from what I remember of his father, he would be admirably fitted for California. Gambling is, you may say, the prevailing future of business here. The mines are a lottery you may work for weeks and not make a dollar a day . . . A reckless devil-may-care feature is discernable in every department of business, and I can show you more gold and silver in a single night than you ever saw before piled on the Monte tables of the gamblers. On two sides of our public square every house is devoted exclusively (almost) to this business, and you could not but be surprised at the crowded state of the rooms and the public manner in which it is carried on. Every gambler is a gentleman here.

Such being the state of things here, every person reckless of their money, almost any business pushed with steadiness, close attention, and perseverance must in 4 or 5 years yield independence.

John Lash,[31] what shall I say to him? I would like to give him the same advice as I did Brother John, but I think he will forgive me if things should turn out bad—he is not married—a very material difference. The first money Hosea and I get we will send it him, and we will go hand in hand together through California and the West if he chooses. Oh how often we have thought of him since our separation, and wish for his song and dance, and just until we got quite blue over it. Nothing could give me more joy than to meet him at the landing of San Francisco, should that ever happen (God grant it may) we will give him a welcome that will only be equaled by the one, when we get home again. Then if we only can make our "eternal fortune"—but I am anticipating. John, come out if you can. We will tempt you the first moment we can raise the "dinero"—depend on that.

It is not unlikely that Hosea and I should go into the gardening business, can we find a good location and a convenient market, we will think about it at any rate.

Hosea, in his letter states our intentions concerning the company. That I have been treated bad by Captain Taylor there is no doubt. He quarreled with me and abused me, and then left me here with all the business, debts, sick, (his own brother included) and trouble on my shoulders. I have written to him reviewing the whole ground of quarrel, and speaking with plainness, candor and courtesy, and will have an opportunity of sending it him by a member next week. If he have manliness enough to retract, all differences shall be buried and forgotten for nothing grieved me in all my life more, since its occurrence than does it. That he mistakes, and has mistaken me, all along, is most certain, probably he will see clearer now. Charles, his brother and I, to the last, were on the best of terms—he never knew the particulars of the affair, and I requested him not to ask, that nothing might spring up between us but good will and fellowship. The Captain's answer shall decide the matter for I shall lay it before the stockholders, with my letter and his reply. After I send up the letter I will transmit a copy to you.

My course is this: At the end of two years I will have . . . all the money I need . . . and I will make the best use of it I am capable of. I shall abide by the agreement made with them, so far as the division of the earnings is concerned, and if nothing is realized from the other members, or should it fall short of the original capital, and my earnings during the two years can cover the deficiency, they are welcome to it. This extends over a space of <u>two</u> years from the time I was left in charge of affairs here by Captain Taylor; beyond that time I am free of all obligations to them. Hosea sustains me in this course, and will act for them on the same conditions—but before I lay this before the stockholders, I wish to hear from Captain Taylor and therefore postpone the matter until next steamer: It may interest you, therefore I mention it. I know not whether you will approve of it or not—but it is the best course I can see at present.

But a truce to all such talk. You know not with what anxiety we have been waiting the arrival of a mail steamer. Three mails, as I said, are now due, and I noticed as I came up street about an hour ago a call for a meeting on the subject, a friendly one may be expected. For the Californians, as you will perceive by the Constitutions I send [are] you a pretty saucy set. At the Democratic meeting mentioned (I know not what the paper says not having time to read it) the ball was started with a regular barnburner push[32]—please tell Dr. Mason so. We are all right out here on the main principles, ... How are all the family? I dare not think too much about you for fear of the blues. But it will all come right some day. 'Tish shall have her piano and 'Vine her shawl. And when we come home we will buy about a dozen fine Mexican blankets—they will make splendid bed coverlids. Mother, Emma, 'Vine, 'Tish and Warren, God bless you all! Uncle Pinckney and Aunt Elizabeth, too. Oh! Won't we turn the house upside down when we get home again, John, Joe, Hen, all, all, are remembered with the strongest love.

Justice to my business requires me to stop, I have yet a few short letters yet to write.

More than once, I would have sacrificed a year of my life for five minute's advice from you since I left Reading. I sincerely hope and trust Hosea's and my course here will meet with your approbation—God grant that it may yet be our lot (yes, particularly mine, for I have already caused you so much trouble)—one happy lot, to guild your older day with happiness, and joy and plenty.

Farewell! Affectionately and truly your son,

E A Grosh

∽ Letter from Allen Grosh:

November 10, 1849
San Francisco
Dear Father,

This will be hand you by Mr. Wineland of Easton, Pennsylvania an associate and companion during our journey to this country.[33] He returns by sailing vessel and will probably be a long time on the road. I therefore send but little word, referring you to him for California news, general and particulars, which he can give with greater interest to you than I can write it. I recommend him, my dear father, to your friendship as one every way worthy of it.

We have had wet weather. . . . We are still "dwellers in a tent" and <u>such</u> a tent! Medium horsehair sieve cloth would yield better protection. Next Monday or Tuesday I hope we will be able to provide better quarters for the sick at least.

I received a few lines this morning from Uriah (the first word since they went up) It is dated "Jacksonville Oct 27th." Abbott has left them and is at work on a bar in the river[34]—Zerbe has been sick—but is well again. They have made about $3 each per day, and the ground growing richer as they work it.

Hosea and the rest of the sick improve, slowly when the weather is bad, rapidly when it is good. Day before yesterday Hosea was all over town with me.

Yours of 6th August was received last week and I will answer it by a gentleman going in the steamer on the 15th.

In good spirits despite of mud and rain.

Truly and affectionately,
Your son,
E. A. Grosh

∽ Letter from Allen Grosh:

November 30, 1849
San Francisco
Dear Father,

We received yours of August 1st about the 1st of this month and I must say that your complaint of our not having written surprises us much. Hosea wrote at Tepic, and sent by the U.S. Consul, who left for the states while we were there and I wrote at San Blas, and confided in the courtesy of a stranger, per necessity. Hosea wrote at my particular request as I was then too unwell, and I wrote at San Blas to let you know that I was able to do so. We are the most sorry for this as we did not get to see Simon Seyfert when he left and you had to wait until the next steamer before you heard from us. I would here remark that we did not write by the last steamer (15th November) owing to a mistake as to the time.

. . . Uncle Pinckney's difficulties, but above all mother's failing health were all matters of so serious a nature that they were constantly before us. God grant that the same kind Providence which has so long protected your family from the ravages of death may still preserve us, and may the separation of three thousand miles of desert, plains and mountains but draw the closer those bonds of love and kindred, and find them ten-fold stronger when again we meet, on the Atlantic or the Pacific.

Your pecuniary troubles, father, I regard but little and if you can only "hold on" at something until next spring I feel certain we can lend you at least a little aid.

We have been unfortunate here so far, but I think we can hope for better things in future. Dr. Martin's sickness prevented us from accepting a situation in the *Pacific News*[35] Office at $50 per week, and Hosea's illness also deprived me (and perhaps him also) of a situation in the post office at $150 per month. Either situation would have enabled us to lay by about $100 per month. On Hosea's recovery I had a very severe attack of the piles, which threw my back again since that time. The roads to the mines have been rendered impassable, and thrown so many persons into town that a situation has been hard to get. A few days ago I accepted an offer to go up to Sacramento City as clerk in a grocery store about to be started there by Farrelly of our party and a young man who came with us across the country by the name of Davis[36] of Philadelphia were "chums" during the journey . . . While I am working, Farrelly has made Hosea an offer, which will probably be well for both parties. Farrelly is to build another house, and provide a small stock of groceries, etc. Hosea is to attend it for half. He will have a stand for the sale of fruits, nuts, cakes, hot coffee, tea. Hosea and I have just suggested chocolate, Mexican fashion, which is to be had in no place in town. It is a splendid beverage and easily made. Groceries yield about 100 per hundred and the articles in the store would yield twice that at least. Remember California is a great country. Hosea, Farrelly, and myself are heartily sick of our company and this arrangement will relieve us of our boorish, hoggish companions, and enable us to spend the winter in a pleasant and agreeable manner.

So cheer up, father, and rest assured that both you and Pinckney will be soon relieved of all the troubles that now encumber and annoy you.

Mr. Davis is the gentleman that suggested the possibility on which is based my "gold washer," and we are going to build a model (half size) 3 foot long and 1 foot in diameter as soon as we can get at it at Sacramento City and try it. We propose as a test to mix up one ounce of iron filings with two bushels of dirt and if we can succeed in extracting nine tenths of the iron, will call it complete. If not, we will try other plans. Mr. Davis has been working in the mines for the last three months and he is very sanguine of success. . . . You express willingness to come out here if we can find a good comfortable home for you. Thank you dear father, a thousand thanks. If in the course of a year or two, such a place can be found, possessing a perpetual spring-like climate, good land, plenty of water, fine scenery, healthy and. . . . with a good market, we will secure it, and urge you to come out. But San Francisco and the Sacramento would not suit both in healthy and disagreeable. California I think will be our future home and what would I not give to have you and Uncles Pinckney, Frank, and even Bayard (is it impossible?) out here—but not before a comfortable home is provided. Things can't go here as they have done, and are doing, this spirit of restless expectations, which pervades all business must have an end, and where that end comes, I want to be provided for it by having my resources converted into ready money—then I will invest in land and not before. California, you may depend upon it will be the scene of a revolution never before equaled, one of these days.

Next spring we try the mines, and I am going into that business just as I would go into any other, commence small, and as I acquire the means, extend my operations, always using the best means within reach. John Lash would be of inestimable value as a companion and partner in such operations, and if we can possibly raise the means, he will be out here by next spring. We are very anxious to hear how the trial terminated, but I do not think Newkirk can touch him . . .

Congratulate grandfather on his marriage, and tell him that we should have done so before, only that we thought we possibly might go to the mines, and dig enough gold for a ring for grandmother before this time as we were anxious to send her our good wishes, in this manner, out of the first gold we should gather. But, alas, we can only send the promise to do so, and we will redeem it at the earliest possible day.

We have had nothing of the cholera as yet in California and I think the geological structure of the country will prevent it prevailing here to a great extent. I have however heeded your advice, and shall follow your advice as to its treatment, should it appear. The crock of conserve of spices which was sent around the Horn came to hand in excellent order and has been of much benefit to Hosea.

(Second Sheet)

Daniel Byerley's sickness, and D. J. Werners' accident, were mentioned to the party and all seemed to sympathize with them. We were very sorry to hear of Henry Lash and Mr. Deahl's loss.

Samuel Klapp is at Sacramento City, where I am going. His last request, when he started up, was that I would mention to you that in the war against you carried on by C. R. [Christian Ritter] and L. B. [Lewis Briner],[37] he was, unwittingly, induced to play a part against you. We had a long talk about it, and I have not the least doubt that he was used as a tool by them, and that about all he said against you and me, too, was said under the influence of passion, excited either by circumstances, or them. He blamed himself very much for not having called on you before we left Reading. He also gave me an explanation of the course he took at the society meeting . . . so impotently. He says . . . <u>wrongfully</u> and accuses L. B. of having shown it him for the purpose of exciting his anger against you. He seems truly and sincerely sorrowful for the part he had taken against you. But of this another time. Him and I are on the best of terms.

Aunt Letitia's school is a failure—well I am glad it is no fault of hers, and I hope the time is not far distant when her brothers will have the pleasure of enabling her and 'Vine both to acquire an education commensurate with their desires. It makes me sad, sometimes, to think that when [I] again see them they will no longer be mere girls, and that their giddy pranks will no longer either vex or amuse me. And Emma going to visit you! Would that we all could gather around the old family altar of happiness this fall! But no! It must not be and perhaps it is as well. But when we do meet won't we have a time! Tell Emma to kiss the baby for me.

I will write to John by next steamer, and would have done so with this only we have no accommodations for writing—being obliged to do it all on the knee.

Uriah Green, and William Zerbe, we expect down daily, they and Tom Taylor [and] Abbott went to the mines together. Abbott left the party. Captain came down to attend to business and Uriah and Zerbe remained calculating to winter [there]—but the rainy season set in too early and they were obliged to leave. They were well. Farrelly, Hahs, Axe (who wishes to be remembered) and Witman are here in our tent, with us. Rapp, we heard from a few days ago, he is doing very well up the northern mines and is well.

The Reading party attached to Gordon's party went up to the mines. . . . Doughton was sick up at the mines . . . he would winter it out and they left him gathering <u>acorns</u> to assist him in his resolution. Turnbull, Fisher, Deem,

Bowman and Shawman are here and look in <u>excellent</u> <u>health</u>, and they say they never enjoyed better in Reading—they have had hard times of it. They will try it again next spring.

Turnbull and Shawman were going right-straight home when they first came down, completely sick of California, mines and everything else—they even shipped for Panama. But their companions induced them to take the "sober second thought" and remain.

Johnston S. Flack is obliged to go home. His is indeed a hard lot. He has been unfit for work ever since he has been here, and goes home with a great deal of reluctance.

We hope mother is in better health, and tell her that we often think of her out here. Every Sunday evening I devote to home and its memories, and oh! Who can tell with what rapture each face is called up by memory—but I must not think too much on these things now.

As this will reach you about New Years Hosea and Allen send you a "Happy New Year!" Yes, thousands of miles we send it—will the distance weaken it? No—oh! no.

Give our love to the girls, Warren, Aunt Siggie, Uncle Pinckney, Boas's, Wilsons, and don't forget John Lash . . . Nettie and Charlie too—and . . . not gold dust to send we can at least send kind wishes, and exchange old affections.

Truly and affectionately your son,
E. Allen Grosh

   ∞ *Letter from Allen Grosh with addendum by Hosea Grosh:*

December 1, 1849
San Francisco, California
Dear Father,

I last evening deposited in the post office an answer to yours of August 1st, but as several things occurred to me since I again write and send by Mr. Johnston S. Flack, of our company, who returns on account of bad health. He is probably the last that will return from that cause, as the rest of the party, (Hosea not excepted) are in good health and seem thoroughly acclimated.

Somehow or other I forgot to mention that Henry Kerper, returned by sailing vessel on the 15th of last month. He goes to Panama in a sailing vessel, and to New York by steamer. He returned much against his will—but he had no choice—his health absolutely required it. Flack returns from the same cause, and as reluctantly.

I would also mention, for fear I might forget it, that Mr. Jolly [has] . . . family is in Reading having moved there last spring. Mr. Jolly is a fine young man, and one in whom I am much interested. He was taken sick here, and was obliged to raise money at an awful share (to you at home who are not used to such things) of 15 per cent by giving a draft on home for four hundred dollars. The amount thus raised just carried him through a couple of weeks at one of our hospitals, at $10 per day!—the usual rate—and he is now able to take care of himself, and earning about enough to pay his board. He is an excellent clerk, and brings with him letters from Hon. J. Freedley, M.C. from Montgomery County,[38] and others equally good and yet cannot get a situation to do his best, so crowded in the city at present, by a class entirely dependent on their labors for sustenance. I think this will correct itself before long. There is too much money in the country to fear a pressure, and the very crowd that is now clamorous for something to do will create a market which will call in the assistance of others, and before spring wages will be higher than ever. Carpenters are in demand at $12 per day—who talks of Congressman's wages now?[39]— and there is enough to do about town to employ half as many more as now are at work.

I also forgot to mention the circulation of a rumor of a very painful character. It is that the steamer due here the 15th November and not yet arrived, has been lost at Point Concepcion about 200 miles down the coast, and only 20 passengers saved, and most persons, I am sorry to say, are inclined to believe it to be correct.[40] Something must have happened [to] her, and as most all communications up the coast (the steady northwest winds rendering a sailing vessel's passage very tedious) is conducted by means of this line of steamers, it is highly probable. It is terrible to think of. They hug the coast too closely at any rate, . . . In one case we came very near running into land, and we were not more than fifty yards if that from the breakers, when their white tops were first visible through the thick fog, which envelopes this coast nine months in the year.

Had Hosea recovered earlier, I think we would have went into the gardening business. The rainy season commenced about 4 weeks ago, and everything is as green as a meadow in spring time. Every foot of California, capable of cultivation can be brought into use for one crop a year, and when the means of irrigation exists, two crops can easily be raised. What will Uncle Pinckney say to the price of vegetables here! Potatoes 35¢ per lb.—onions $1.25 ditto— sweet potato squashes from $2 to $5 a piece, cabbage, turnips, radishes, hardly to be got, and the only limit to the price the conscience of the seller and the purse of the buyer. Poultry business would pay here. Eggs <u>only</u> $6 per dozen when they can be had.

For this past week we have had splendid weather—cold clear nights and warm sunny days, and indeed I am quite reconciled to my home. The eternal veil of mist which shrouds everything during the greater part of the year is now lifted up, revealing a country rich in beauty and poetry. The name "Italy of America" of Frémont[41] is no fiction. The scenery around San Francisco is just what I imagine that of Italy to be. The atmosphere is so pure, the sky so bright and soft, outlines of the surrounding hills so chaste, and beautiful in form, that you unconsciously associate it with classic land. Can you imagine Indian summer, with green bushes and hills? The atmosphere with all its softness but without its hazy indistinctness? If you can you have San Francisco as it now exists.

. . .

The rainy season as I before remarked, set in about four weeks ago. The present spell of fair weather is expected to continue until the second week in December and then <u>the</u> rain begins. If San Francisco does not float off into the bay it will not be for want of "easy communication." The rains that we have had rendered some of the streets almost impassable. I saw, the other day, a wagon <u>float</u>—that is, the box was setting in the mud, with the wheels raised off the bottom. Stockton, they say is as much worse than San Francisco as San Francisco is worse than New York. Bowman, of Reading, who came down yesterday says the mud in all the streets is more than knee deep, and some places utterly impassable. Sacramento City, I expect, is not much better. Thick boots—of the commonest kind—are worth $32 and <u>good</u> thick heavy winter boots are worth from 3 to 5 ounces [of gold], even more. Hosea was fortunate two pair of his came around the "Horn"—mine were lost—of course he divided with me—we have adopted the old custom of having everything in common.

I wonder if it would be possible for Uncle Moses to make a pair of canvas soled, quilted bottom boots of the very best quality, and send them out here instantly. I would designate a place in San Francisco where they could be left. I would willingly give 4 oz. for 2 pair for working in the mines next fall and summer. I will write to him about it.

Hosea and I are in excellent spirits and are going to make our "eternal fortunes" right away though we now haven't got a cent to commence with. I do not go up to Sacramento City until Sunday (day after tomorrow).

Give our love to the girls, Uncle Pinckney's [family], grandfathers and the rest, and tell them that we often think of them all. Hosea is writing to you. We both send by Mr. Flack, who will hand them to you, and give you any information he can with pleasure. Remember us to all our friends, and tell John that we will not rest contented until he is out here. Tell mother that we will send her all the flower seeds we can find that is worth the planting. The California poppy is a beautiful flower.

Adieu until another steamer,
Truly your son, E. A. Grosh

We are glad you mentioned th . . . tteson are out here—we will try and find them out.

We have not . . . if we find him will attend to that matter, and remit immediately.

Per politeness of Mr. J. S. Flack

. . . B. Grosh

. . . Reading

Pennsylvania

I write what here follows for your eyes only.

There are seven in our tent at present viz. Hahs, Witman, Axe, Flack, Farrelly, Allen, and myself. I wish to say a word in regard to their dispositions, etc. Hahs is a boorish Dutchman of the regular Berks County stamp. He is not particularly good natured, though not perhaps in that class what would be called crabbed; nevertheless something of a fault finder; in fact, I think the words in italics express his character more fully than any others could do adding that he even beyond the majority of that class is ignorant and careless of everything appertaining to the gentleman. Witman, is different, very different, he is witty and displays no little talent in telling a story in which his imitations are excellent, to the life, he has the art of bringing out all those things in life to make him a pleasant fellow . . . but . . . he is bound to take his full share when helping himself, let who will go without, or more, as much as he wants though it be all: and then he is not over honest and will take what belongs to others without asking leave should he feel any need of it and in some cases to sell and thus to raise money for his own use when short. Axe is good natured, honest and good hearted, but extremely close, ill informed and dull of comprehension though still his good disposition makes him by no means as unpleasant to live with as the others before described. Flack who leaves us tomorrow has more mind than any of the preceding and in rough parlance "is no fool" but he is an habitual grumbler and has been further soured by long sickness and ill health being now obliged to return home on that account. For he is unable to work though he could if able have had constant employment at $12 per day.

Farrelly is in character and manners a gentleman and in nearly every respect as pleasant a person to live or be with as one could wish.

You can judge that such a crowd would be by no means a pleasant family to live in and will I know rejoice as I do that there is a fair prospect of my being free from such a fate. Should the plan mentioned by Allen in his letter by steamer not turn out anything. Farrelly who is as sick of the others as myself has leased a lot and will in all probability build a shanty on it that will be a comfortable place for the winter and then perhaps Allen may be able to procure me a situation in Sacramento City even if all these fail I do not despair of getting employment here, which shall include board and lodging.

I would have written by the steamer but felt a sort of horror, almost, at sitting down in the tent and taking my paper on my knee to write more than I could avoid so I left Allen to speak for us. . . . Your affectionate son,

Hosea B. Grosh

December 1, 1849

The steamer arrived yesterday with some 500 passengers leaving about the same number on the isthmus. I have written to John G. Jones, December 2 1849.

❧ *Letter from Hosea Grosh:*

December 1, 1849
San Francisco, Northern California
Dear John,

I had intended to write long ago but I have been too busy or too lazy to write at all except a few lines by the steamer of November 1st just as I had fairly begun to recover from the dysentery. I did intend to send you a full detailed account of our travels by sea and land which I have not the slightest doubt would have been vastly interesting, throwing those of Mungo Park, Gulliver[42] and all former travellers in the shade in that particular, however from laziness this has been neglected until now and now it is too late for the present as the person by whom I send this will in all likelihood go aboard tomorrow so that I will finish this tonight. I will send you an outline in this and prepare for a future time a full and particular account giving our route; the rivers, mountains, valleys and plains we crossed and the cities along our route over Mexico. We were a number of days at Philadelphia and were as many more going down the river and two at the breakwater on account of bad weather outside and finally at sea on the 4th of March for Tampico which we reached after some 20 days . . . sail having met with a storm in the gulf after reach Tampico . . . being within sight of the harbor when we were obliged to put off to save ourselves from shipwreck by which we lost a couple of days. At Tampico we were detained for 15 days which probably laid the foundation of much of the sickness after that we were afflicted with on our way through Mexico. Almost the first day out Allen was taken unwell which hung onto him until after we reached San Luis Potosi. A couple of days before we reached that place we lost one of our men of congestive fever. We were detained a good many days through sickness in the . . . At San Luis Potosi the government seized the mules that we had hired to carry our baggage for carrying Army baggage this detained us ten days and I don't know when we would have got off but we bought two wagons and sta . . . but we found that when we got on the road that the wagons were overloaded so that we were obliged to lay by a day till we procured a dozen jackasses that could carry about half the baggage so as to leave the wagons about well loaded and not too heavily loaded so then we moved on over all kinds of roads; hard as rock and smooth as a floor or perhaps rough or sandy sometimes on a trot sometimes all hands off and to the wheels to get the wagons through. We passed through Blados, San Juan, La Patlan, Sapatlan, San Miguel and some other cities and towns besides haciendas and ranches without number before reaching Guadalajara. On the way one of the servants stole a couple of horses and West and Farrelly (one of our party) went back to try to recover them. We sent one of them off as we detected him in a place for stealing a couple of horses and tied him for the night and packed him off in the morning. When we reached Guadalajara we found all the funds were gone. After some deliberation we advanced among hands $250 taking a bill of sale on the horses as security. This enabled [us to] pay our debts and provide for our progress toward our destination the great bugbear at Guadalajara was the . . . which it was said that we could not take our wagons over without taking them . . . took them down by hand safely we pushed on over road were rough almost enough to break almost any wagon you could raise up and over steep little hills along the creek and stopped at a place bearing the name . . . whole place through which we had come and had a day's journey to go to get out of yet. Next morning we started over a most splendidly paved road, of which better is scarcely to be found anywhere than in Mexico, for about ½ of the hill which we had yet to pass then came the "try of war" for the rest of the [road] was not only very steep but covered with loose stones, so that we had even in some places to take out the mules and draw the wagons up by hand as the mules were unable to get foothold suffi-cient to draw. When we arrived at the top of the mountain we went down the other side over one of the roughest roads I had ever seen. Thus we got over the barrancas so much dreaded making in Tepic two days about 54 miles and arriving each night completely worn out. Tepic was the next great point in our route. This place is situated some 60 miles from the sea coast so as to be clear from mosquitoes, sand flies, and fevers which infest the nearest coast; it does all the business or at least all the inland business of San Blas, the coast town which is a miserable little place situated among salt marshes and stagnant waters which breed myriads of mosquitoes and sand flies. Tepic is a very pretty place and the general impression is pleasanter than any Mexican city that I have been in. We remained here from about the 12 of June to the 21st when we started for San Blas which we calculated to reach by the 24th in time for the steamer which was then expected about that time. The road was awful. We had one upset the 1st day when one of the wagons containing the greatest portion of the baggage and two persons fell onto me crushing me to the earth and then rolling over me down into a creek bed down the bank of 8 or 10 feet leaving me again free though stunned partially and crushed so as to be scarcely able to walk. After a short time however I was able not only to walk but after a short nap, the effect in part of the accident and in part of some port wine that I drank at that time, while the others were getting this wagon up and the other across a bad gully I woke and helped carry the last of the baggage and followed the wagon on foot during the remainder of the day. . . . The next day we went along very well and stopped at a place at which we got our first taste of sand flies and mosquitoes and a sweet taste it was. It commenced raining during the night and tired as we all were scarcely a soul got a wink of sleep. For myself I walked about with my gum blanket[43] on all night. The mosquitoes could not have kept me awake but the sand fly's bite resembles in feeling that of a hot point of irons touching the skin and this capped the climax. The next morning we started about 10 o'clock the rain stopping about that time and floundered through the heavy wet roads having a clayey mud of several inches in depth which was extremely tenacious making it unpleasant for all on foot especially those with the wagon. Just before reaching San Blas we came to water some 2 feet deep through which those who went before us had to wade though I being with the wagons rode getting in at one place the wheel ran over my foot which disabled me from walking this I did not get fully well of until after I got on the vessel so as not to be obliged to use it. At San Blas we established ourselves in the upper town on a high steep hill when we arrived after dark on the 24th of June at this place. We remained in a deserted monastery, hiring a room at 25 cents per day and occupying four of them at least before we left, until the 11th of July when we set sail in the bark *Olga* of Boston for San Francisco. I say we though all of us were not there. Captain Taylor, Samuel Klapp, and one other set out for Mazatlan about a week before with the mules hoping to find a better market for them at that place; in this expectation they were disappointed as we afterwards found. One of the company remained at San Blas on board a vessel on which he had engaged to work his passage as carpenter. Two, having some funds of their own by aid of which in addition to their proportion of the cash on hand they were enabled to get tickets for the steamer they went in her. Six of our men insisted on drawing their share of the funds and going on board a craft lying in the harbor and up for San Francisco which they did and by what seemed to be a special providence (for she was scarcely seaworthy, her provisions poor, her water worse and scanty capable of carrying not more than sixty passengers even as they stow vessels on this side of the continent she had nearly double that number) she was struck by lightning the night before she intended to sail. . . . in many respects but we found . . . that we fared rather better than most steerage passengers . . . Off Monterey we lay 5 days making and losing ground alternately . . . fogs and strong currents before we were able to get in. There we procured fresh beef which was a treat indeed it being extremely fat and tender as nearly all the beef of this country is, except what we get at San Francisco. We lay at Monterey for a couple of days, which is handsomely situated and is quite a pretty place with a splendid climate, from there we came here with a light but fair wind in about 40 hours sail and the first trial made it through the mouth of the bay we were soon in San Francisco harbor. I have not room or time for anything more than to say that here we lost Dr. Martin by dysentery and on the same day that we buried him, September 14th, I was taken with the same complaint and for eight weeks was unable to [walk] down street more than a square and back. In relation to coming out here I would advise that you come not out unless you have capital of your own sufficient to give a fair start or until we can have it for you. Then we have dysentery and the Lord only knows what other diseases which are very fatal to the unacclimated scarcely less so than would be cholera itself. Your business as mason is not good as yet though the time must come when it will be. Baking is very good if you have the means to carry . . . on. Allen will not get off till Tuesday yesterday the steamer arrived with 500 passengers leaving nearly as many on the isthmus though what they will all do this winter I can't even guess; others are going almost as plentiful as the . . . . My health is good now having passed from me entirely.

Your brother,
Hosea . . .

∽ *Letter from Hosea Grosh:*

December 30, 1849
San Francisco
Dear Father,

I will not probably write a very connected letter this time; however, such as it will be, write I will anyhow. We are now in a house, and are comfortably situated, but before we got into it our discomforts were not inconsiderable, especially when it rained. Before the last spell it had rained only for a couple of days at a time and then cleared off so that we had a chance to dry our blankets and other things that were wet in fact scarcely anything kept dry that was not in trunks or gun bags. The last spell was different, it commenced raining on Tuesday evening and rained steadily until Thursday night. Friday was so showery that we had no chance to dry anything. Yesterday the sun shone out clear for a short time in the forenoon. I hung out the blankets to dry, the rest were at work on the house, but scarcely had I finished hanging them than it commenced to shower once again. I was obliged to take them in again wet as they were which could happened . . . at the using some sticks. I succeeded in fastening it again at the whole wall was blown loose from the ground. At night we went to taverns and lodging places to lodge it being impossible to sleep in the tent everything being soaking wet, that was not covered by the gum blankets and they were damp. The bottom of the tent was one bed of mud. The gum tents are no great shakes, they let through the water like a sieve, and they are rotten besides; a good canvas one would have been much better. In addition to the heavy rain and wind it was very cold thus increasing our discomforts. The next day we moved into the house which though the floor was all wet was far more comfortable than the tent. Though this exposure was not felt particularly at the time it was the means of bringing on a slight attack of diarrhea of which I am just getting better. Allen is at Sacramento City. I have heard from him several times since he has been up. He is well satisfied with his situation. He has tried though in vain so far to procure me employment there also, says that he thinks that I will like the place if he can get me there of which he has no little hope. My own situation is not as unpleasant as I anticipated but nevertheless I shall change it as soon as possible. My personal conditions, bodily I mean, are all I wish though I still in other aspects do not feel perfectly contented. I shall make application for regular employment as soon as I am perfectly restored from the slight attack I have just passed through. We have had a large fire here which has destroyed nearly a whole block of buildings and a large amount of property.[44] The buildings are going up again rapidly though they are intended only as temporary ones it being the intention I understand of the holders of the property to put up brick buildings as soon as materials can be procured which will hardly be before the middle of next summer. The weather for a week or so past has been very pleasant far more so than when we first came here then every afternoon it blew almost a gale from the northwest. Now the winds are light. Then the weather was exceedingly cold calling for heavy overcast . . . weeks ago saw snow on the hills on the other side of the bay here we have had none and I am told that is very seldom seen here, last winter but once, and that once surprised the natives. I find that Frémont's remark that the warm pleasant weather of the winter contrast pleasantly with the cold chilling winds and foggy days of summer is strictly true. Mechanics here are getting to be troubled with the Oregon fever, that country from all accounts being vastly better for their pockets and also the living is better and in every respect, life there must [be] pleasanter than here, from $10 to $15 per day found, being the run of wages for carpenters and everything they most need cheap laboring men $8 per day found and so on. So things go. Captain Taylor had returned from the mines intending when he first came down to try and get up provisions for his party at the mines but they being convinced that the thing was impracticable sent him word to that effect and come down with exception of Abbott who had been ordered from the party by Taylor a quarrel having occurred between them. There are now eight in this place six living in the house at present Farrelly, Axe, Witman, Zerbe, Hahs, and myself. Taylor and Green are boarding. Green has gone into business here in a store with Edward Riggs each furnishing I understand about $1500 capital. Taylor talks of going to Oregon just now but he talks of many things by turns so I do not know whether anything will come of it or not. Abbott is on the Tuolumne (I believe is the stream) one of the tributaries of the San Joaquin. Rapp is on the Yuba one of the tributaries of the Sacramento. Allen and Klapp are in Sacramento City. Klapp working at his trade there though at last accounts from there he had been unwell though . . . Allen's health he tells me nothing about . . . there has been a great fire at Stockton the centering point of southern mines which has destroyed nearly the whole business portion of that place.[45] We have had several shooting scrapes here in fact they are getting quite common, the Council have offered a reward for the arrest of "Withers" who stabbed a man and killed him of $3000. Gambling is going on as usual though the fire burnt out several of its principal houses. They will be up again before next month is half over. It would seem that anytime after next midsummer the masons trade will be good. The other party from Reading have lost their leading man Doughton[46] who died in the mines, I believe of dysentery six are in this city at work, Diehm, Stebbens, Turnbull, Fisher, Scharman, Bowman, I mention so that if anyone should inquire of them you can give some information. Stillwell and Stroup are in the mines and doing well at last accounts. Bitting is, I believe, also there at all events at last accounts was well.

I remain your affectionate son,

Hosea B. Grosh

I think that I have forgotten something which I intended to write . . . a U. Green is going to send this by a friend. I must close it now. I will add however that Captain Taylor has recovered from an attack of fever and Witman from a severe cold which deprived him of his voice for a couple of days. The most of the provisions . . . around were sold on what terms I do not know, everything being in Taylor's hands but I believe to some advantage.

1850

Letter from Hosea Grosh:

January 5, 1850
San Francisco

Captain Taylor has left here for the Sandwich Islands[1] so that the only portion of our company that at all deserved the character of a company working together in California is now broken up [and] separated. Farrelly, Axe, Hahs, Witman, and myself are here in the house. Zerbe is also with us. Green has gone into business with Riggs. Abbott is still in the mines of the San Joaquin. Allen is in Sacramento City and Klapp is at work at his trade in that city. Peter Rapp is in the mines on Yuba one of the tributaries of Sacramento. We have heard that he has done very well! This is the whereabouts of all the remaining twelve as a working company in California. We are broken up. Each member will make his "pile"[2] in his own way and account personally with the stockholders. On our arrival here we had no funds but raised a thousand dollars on a draft for $1200 from home giving the bill of lading of goods and provisions as security for its payment. This was not done until we had been more than two weeks here. . . . Taylor went to the mines taking with him Green, Abbott, and Zerbe, having first kicked up a quarrel with Allen, commencing about a blanket, and ending by Taylor forbidding Allen to seek to join them at the mines. He very coolly told Farrelly, with whom he was displeased, because he thought that working for $12 per day was better than going to the mines, that the company with him was large enough. He left Flack and myself sick, Allen and Farrelly also with, after the doctor bill was paid, about $275 with which to support us and pay all future doctor bills etc. taking with him in the neighborhood of $500. I find by a letter written by Abbott and Zerbe while they lay at Stockton that there they (i.e. the whole party) intended for a while to give up all thought of going to the mines and were about to settle for 3 or 4 years on a piece of land that they were going to garden. But that failing Taylor proposed breaking up and each one going on his own hook and talked according to their account very insultingly to those who opposed his scheme. I would here remark that Taylor while coming through Mexico was very arbitrary, obstinate and to anyone that differed from him at times very insulting. This set most of the men against him though the officers after a thing was done and there was no help for it feeling desirous of keeping the company together and having as little fuss as possible generally supported him unless he actually went beyond all reasonable bounds when little or no good resulted from their opposition Taylor generally taking the responsibility and transacting the business without consulting them. After they got to the mines Abbott and Taylor had a dispute concerning the disposal of the gold gathered which resulted in Taylor ordering Abbott from the party forbidding him to eat or work under the same roof or in the same company with himself. About the time the rainy season set in Taylor came down intending to take provisions to the mines but this proving impracticable Zerbe and Green came down here. Before they went up it had been arranged that Taylor should leave an order which would enable Allen to transact the business as well as if Taylor himself were present, as he alone was recognized as agent or owner of the goods by the consignees of the *Susan G. Owens*[3] there being granted to him alone . . . in procuring even our private effects and putting anything farther out of the question though some things might have been sold to better advantage than after Taylor came down. When Taylor came he sold all but a barrel of beef, a couple of pork, and flour, etc. Since we have been in California and while coming through Mexico Taylor treated the officers as if they were only appointed to save him labor and not to have any say in directing affairs. Even the stockholders though they have expected them to do their duty have never in any way acknowledged them as such. I have stated these things as they will serve to make many things clearer than without, though it is too painful to enlarge upon, to say nothing of the imputation that one may be obliged to lie under of spite or prejudice. I have made the statement as correctly as possible drawing most things from my own knowledge or Taylor's account though I have drawn some from the account of Zerbe and Allen . . . [*end of page; balance of letter missing*]

∝ *Letter from Allen Grosh:*

January 8, 1850
Sacramento City, California
Dear Father,

Tomorrow I start for the mines. Jackson's Creek,[4] 65 miles from here, and I am sorry to so say without Hosea. I have now been in this "city in the woods" for more than one month and during that time have not heard one word from my brother. How he is getting along I cannot tell—he will probably tell you himself, for this is . . . certainty in the mail between San Francisco and the states—when between here and San Francisco about half the letters never reach the hands of their addressees.[5] The expresses are but little better. I have written to Hosea time and again and quite likely he to me, without us being able to exchange even a word. Is it not enough to try the patience of a saint? For three months, now, I have not received a word from home. I shall be down in about a week again when you shall hear from me. I go up with Mr. Davis, and the store—We calculated to dig and traffic both until the spring sets in when we will start for better "diggings."

When I come down again I shall either go, or send for Hosea so that we again shall be together. I will probably be half owner in the store with Mr. Davis, who every day grows in my esteem. I have but little doubt that I shall do at least more than fair, and will in the course of a month or two be able to send you some aid in a pecuniary point of view. You may depend upon it. Tom Newland is here and doing well . . . He offers to pay one half Pinckney's passage out if he will come. One point at which I am now aiming is to get together sufficient to start with Hosea in the garden business in a month or two. If I can succeed, we will probably open out somewhere on the Sacramento River (by the bye the prettiest river I ever saw) with a patch of 40 or 50 acres. How would Pinckney like to come out? At any rate, [if] I can in any way raise sufficient, in the course of the coming two months, I will send for him, John and Reuben. Hosea and I will then be satisfied. The business will pay there is not the least doubt, with potatoes at $35 per hundredweight, hay $20 per hundredweight, and other vegetables not to be had at any price. Tom Newland wishes to be warmly remembered to you all, not forgetting Aunt Sissie. "If we only had Elizabeth out here," he remarked the other night, "She would set the whole town crazy"—and so she would, for a pretty woman out here is pronounced an angel—with or without . . . and the wretch who would dispute the verdict of the sovereigns of California would very likely dance a hornpipe to the crowd or nothing—and get nothing but a rope for his pains.[6]

Mr. [C. C.] Griffith formerly proprietor of the *Utica Observer* boards at the same house. He is acquainted with Uncle Pinckney and sends compliments too. Tom Griffith, formerly apprenticed in the *Observer's* Office, is also out here. He is an old chum of Hosea's and mine. Ed Blake, Danby's Son, Walter Swift, and several other Utica boys, all old acquaintances, are out here.[7] Besides I know many others whom we met in Mexico. I am perfect contented in California—as much so as I would be in Philadelphia or New York or any where away from you, and though I often think of you all, father, mother, sisters, brother, friends, I am perfectly contented. I long to get to the mines. It is a life, I am fully convinced, will suit me. One thing else I am convinced of—I must make a fortune for us all soon or else I will become so attached to the life of a Californian hunter and prospector that I will be fit for nothing else.

Adieu until I come from the mines. In haste,

Your son, Allen

. . .

⮞ *Letter from Hosea Grosh:*

January 12, 1850
San Francisco
Dear Father,

I have made several attempts to write a letter since I received your letters of September 16th, and November 10 which I was unable to do before the 7th inst. The news they contained though melancholy was not unexpected at least not wholly so for every letter I received I looked through first, with reference to this point; I have had for some time past a presentiment of bad news, but though somewhat expected the stroke was severe; for a time I was unable to read the letter containing the news, the lock of hair having told the tale.[8] I felt that I had lost a mother than whom few were better, kinder, or more dearly loved. I have now but one parent remaining, and my heart is sad for the torn ties. I miss Allen more now than any time since he went to Sacramento City. I wrote to him a couple of days ago and expected to have received an answer before this time but have not as yet. He talked of going to the mines. The man with a home he was intended to take his . . . heard that Sacramento City is overflowed.[9] The street in which their store is situated is overflowed to the depth of 6 feet. This makes us still more anxious than ever to hear some tidings from him. I try to keep my mind cheered but somehow I feel sad and depressed in spirit. One reason doubtless that I do not feel well, having as when I was getting well of dysentery a breaking out, though this is outward and not as before sore from the mouth down apparently through the whole of the intestines. In a few days I will feel better and my spirits recover their usual elasticity. I have but little to write at present though I may have more in a few days. I forgot in my last to mention that I had seen several of my old crony's from Utica. Allen saw one aboard the *Senator* when he went up and he on his return here hunted me out. Allen mentions that he is among a considerable number of Utica folks at Sacramento City mentioning Thomas Newland among the number though he says nothing of how he is getting along but from the manner in which he speaks I should infer that he is doing well. Dr. Osborn is somewhere in California yet, though neither of us have seen him as yet. I opened a letter I had just penned to Allen and sent him a synopsis of the news in your letters; the letters I did not send fearing to lose them, as our mail arrangements here are very poor, as you may judge, when I tell you that not a single one of our letters to Allen had been received by him though they had some of them been written a long time. From Stockton, I received a letter dated Oct. 12th only last Monday though there is a regular line of steamboats running between that and this place. The weather here though most of the time cloudy or rainy is warm scarcely ever deserving the name of cool verifying what Frémont said of the coast region at this place and to the northward that the warm weather of winter contrasted pleasantly with the damp fogs and chilly winds of summer. We every evening hear the frogs singing and everything conspires to remind us of spring at home. The hills are everywhere becoming green. I am assured that in the mines it is beautiful indeed. . . . Americans attempted to force a party of Chileans from their mining place and they in turn took a number of the Americans prisoners and tying their hands driving [them] like they would cattle before them; the Americans succeeded in regaining their freedom and turning on the Chileans hung several of them. What the upshot of the matter will be there is no telling. I tell the story as I heard it, without vouching for its correctness in any particular.[10]

[January] 13th. The sky has cleared, the wind in the northwest pleasant, and bracing weather everything combining to make me feel better in health and spirits; the worst part of the rainy season is past though we shall have rain for a month or so. Yet it will hardly be as constant as for the past month; from the middle of December to the middle of January the rain and bad weather is most constant and the rest of the rainy season which is from November 1st to April 1st has a large proportion of fine weather before the commencement of the dry season. I shall probably be far from San Francisco, for a more unpleasant place to live during what we call the summer can scarcely be imagined. The whereabouts of our men and some causes of the breaking up of the company you will see on a slip of paper enclosed which I wrote before the receipt of your letters. I think that this week I shall go to Sacramento City and if Allen goes to the mines I will go with him. Farrelly has proposed that he and I go, and I gladly accept the offer. All employments here are overstocked with hands and I see no chance of doing anything here, and even if I did, I would rather be with Allen than separated. All our affairs can be talked over between us and we'll no more be troubled with letters miscarrying, etc. Hahs talks strongly of going to Oregon and if he can raise the means conveniently will doubtless do so. I hope that next time I write I will be able to send something more substantial than a letter though this may be my last for a couple of months so you will be in no wise alarmed if you do not hear from me for some time after the receipt of this and I . . . something we should have been long ago had I not unfortunately been taken with dysentery and reduced so low by it. Everything seemed to conspire against us at that time, though doubtless it is all for the best.

Our ideas of the dearness[11] of things are very different here than at home. The scale by which we measure them being generally larger and having a different system of figure. If you at home had to pay 25 cents a pound for common potatoes, 25 cents for fresh beef, 50 for veal, 75 for pork, $2.50 to $4 a quarter for lamb and mutton, 25 cents a loaf for bread (about our 5 cent loaves) they have been as high as 50 butter, $1.50 cheese, $1.00 dried peaches, $.75 to $1.00 apples, the same raisins from $.50 to $1.25, and other dried fruits about the same and all the other necessaries of life in proportion it would seem enormous as it did to us on our first arrival yet in a short time we have become used to it. Wholesale and retail prices are very different. Charges are generally by shillings even in the most trifling things and not as at home by cents.[12] Retailers have to make large profits to pay the rents they do. A small store not near the central business portion of the city paying $250 per month, and some running up to thousands as they approach the central portion rents are so much a month; so are all interest operations and generally everything of the kind is paid in advance. I speak far within bounds when I say that there is not any other city in the world in which property and rents are as high as here. Some frail light houses on a small spot of ground often bringing what in Broadway, New York would be considered an enormous rent for some of their palace-like buildings, some of them 70, 80, and even a hundred dollars a day. I write this to you though I would not have it published therefore, though I assure you it is strictly true. What is to be the result of this enormous inflation? I must say that though a fall is inevitable, the time may be pretty distant. . . . You cannot I suppose obtain a building lot in any portion of the city for less than about $50 per month enormous as this may seem to you who are initiated into San Francisco prices.

∿ Letter from Allen Grosh:

January 15, 1850
Sacramento City
Dear Father,

Since I last wrote you, all my calculations have been knocked into *pi*[13] by the sudden rise of the Sacramento River. We had just got our articles packed for the mines, and sent off only a few hours, when the water was up to our door. While Mr. Davis and I were congratulating ourselves on having got off our goods in time, we were suddenly surprised by the entrance of the water. . . . We had to leave our goods, turn loose his mules, and trust to providence for the safety of both, about half way out to the fort. In passing the first slough (a "slough" is an arm from the river, sometimes cutting up the valley into islands, and at others running back into the country for miles and terminating abruptly.) he had been compelled to unpack, carry over the goods on his back, repack, and continued on until the second slough, when the water had already rose so high as to make it impossible to proceed. He turned out his mules, placed the goods on the highest ground and left them, returning to town in a boat. They are probably all lost, and I expect I am some $600 or $700 in debt.

By midnight the water was over the floor of the store, and we were compelled to take shelter on the counter. Here I remained all the next day and night but was at last compelled to abandon everything to its fate and take refuge with a friend. This is now the fourth day that I have been cooped up in the second story of a house. The water, however, is going down slowly, and by tomorrow, probably, we will be able to get out.

What . . . terra firma again I could say positively if I could only know how Hosea was getting on I would soon arrange matters. It will cost not less than $50 to go down to San Francisco and back, and if I do not have some word by the next steamer I will go down. After that I go to the mines. I could go there now and take some provisions with me, but by going down to San Francisco it will make me as poor as ever. At any rate I go to the mines, and will take Hosea with me if he is not in a good situation.

I write in a hurry and have the blues. Remember me to all my friends and give love to all the family.

Truly your son,
E. Allen Grosh

P.S. I will write before I go to the mines.

EA

(Evening)

2 P.S. Our losses are not so much as we first anticipated—though considerable. I will remain here until next week winding up affairs I will then go to San Francisco when I shall advise you of our future course. I have just returned from the store—not a great deal is lost. . . . Excuse me for my brevity. All will be yet <u>right</u>! Thank God that the debt is incurred in California where a wide field is open to anyone who is the sum of a man.

My dearest love to all.
In haste,
Truly your son,
Allen

I am at present the guest of Thomas W. Cheesman of Philadelphia. Should you meet any person of that name in Philadelphia you may mention me as his chum and warmest friend.

⟋ *Letter from Hosea Grosh:*

February 28, 1850
San Francisco, Northern California
Dear Father,

I did not write, by the last steamer as I had little or no news to send and expected to have been in the mines before this time, however, I am here yet though my next will not be likely to be written here as I think we will be off to the mines in a week or two. Allen came down from Sacramento City early this month sick with dysentery probably caused by taking cold during the flood. He is, however, altogether recovered and looks full as well as usual though not as fat as I do for my face has grown considerably in latitude.

. . . Allen has been at Sacramento City all winter hoping to make enough to take us to the mines but the flood destroyed his expectations. Our going to the mines is, however, provided for and Allen is now at work at a machine which promises to be better—much better—than any now in use in the country besides being portable, which most of the quicksilver machines are not. U. Green is well as are all the rest of the party as far as we know, and we have heard lately of all but P. Rapp. Our boys have gone to the mines already where we will join them soon. <u>By the bye</u> keep the machine quiet. We will write at <u>full</u> length when we get to the mines.

In haste,
Your affectionate son,
Hosea B. Grosh
Remember us to all that know us.

⟋ *Letter from Hosea Grosh:*

March 31, 1850
San Francisco
Dear Father,

We have been disappointed from time to time in not receiving letters from home but yours of February 4 came to hand day before yesterday and fully compensated for all former disappointments, making us feel happy to hear so good an account of your present condition and future prospects. We had read in the papers that you were about to leave Reading and through U. Green that you had done so and felt glad of it hoping that you had thereby done something to increase your happiness. . . . From your letter we see that not only have you left your worst troubles behind but have every prospect of fuller enjoyment in future. You speak of fearing the effect of the bad news on us in deranging our plans and unsettling our minds. Your fears were well founded but still we felt that though one great object of labor for love was gone still with all our sorrow and grief we felt thankful that so many were still left. I must say that for a time it made me feel quite home<u>sick,</u> an effect which my own sickness had not been able to produce. On Allen, the effect was even greater than on me. He had been sick for a couple of days at Sacramento City and when he received my letter he had just arisen from his sick bed. The bad news affected him so much as to lay him up for several days more and when he came down I was astonished to see how thin he looked. But when once together, we seemed to gather strength to endure each from the other. I wrote by last steamer in no little haste and therefore made my letter short. I would scarcely have written but feared that I would have not an opportunity of doing so before going to the mines and thought that you were doubtless anxious to hear from us as much as we were from you. We at that time thought that the rainy season was over but we found ourselves mistaken for we have had considerable rain since and the weather is not yet settled though all accounts agree the rainy season <u>never</u> extends beyond the middle of April. In fact news from the mines reports but little rain there, even the past month though we have had so much here. Allen has had a . . . month of flood in Sacramento City. He had some ugly stumps pulled from his jaws as it could be done with but little pain, they being loosened by the scurvy; rheumatism as usual accompanied it so severe as to draw his leg up slightly though he has it nearly straight again and will soon be entirely well of it. It is well that it broke out here instead [of] waiting till we went to the mines for though but a slight attack and easily managed here, in the mines the case would have been entirely different it being there no trifle being considered hard to cure though I think with care it may be easily avoided. Our town is now on the jump with politics. Tomorrow the election for county officers takes place. For sheriff there are three candidates in the field, Colonel J. J. Bryant the Democratic nominee; Colonel Townes, Whig nominee, and Colonel John C. Hays, the Texan Ranger Independent.[14] Considerable excitement is shown though I must say that I have yet to hear the first public speaker abuse either of the candidates. That sort of thing will not go down here. I am acquainted with none of them except Colonel Jack Hays who I presume in personal appearance disappointed everyone when they saw him for the first time. He is of medium height and slight build, ordinary looking as need be, with a fine eye as ever I saw, quiet, unassuming in his manner, modest almost to bashfulness, but nevertheless liked by all who know him. If you can picture from this a man very plainly dressed and agreeing with the above, you will not be far from the truth as regards the personal appearance of a man who as a ranger stands first and is known by reputation from one end of the union to the other. Bryant the Democratic nominee is a gambler. . . . We are Hays men. From what I see in the papers I conclude that most of the folks on your side of the continent know not how to consider

California politics. The fact is that the political demagogues try to draw party lines where there are naturally none, most Californians being agreed on the most important points and doing pretty much as they please in politics. Our company is pretty well distributed over the country and it is doubtful whether we will meet before we get home. Peter Rapp we have heard of through Reading but know nothing more. Axe, Hahs, Witman, and Zerbe started off late in February for the mines. We have not heard of them since they [left] Sacramento City. Abbott is still where he was. Klapp has, I judge, joined with Axe etc., at all events is in the mines. Green is still in town in business though I believe about to sell out probably to go to the mines. Taylor has returned from the Sandwich Islands having been blown off [course] before he was able to get in his cargo or the vessel even to get ballast, which doubtless saved him from loss as vegetables were higher there than here. As soon as the weather is permanently dry and Allen is perfectly well we will be off to the mines. It is plenty early as yet so we are in no particular hurry though everything is dull in San Francisco at present. Business is at a standstill. Merchants, mechanics are idle or in trouble. We have had one important failure and it is rumored that more may be expected. This will probably be a check but I scarcely think the final crash has come yet things seem to be scarcely ripe for it. I see by your letter an inquiry whether we have met Mr. Heirck yet we have not though to tell the truth it is really very difficult to find anyone in this country unless by accident. In the mines prospects are good and with the machine we will have we hope to do well better than with an ordinary quicksilver machine. It certainly is much more portable [than] a common quicksilver rocker will do. I should say from what I hear, near twice as well as one without quicksilver. . . . I am glad to hear that Joe Wilson and Henry Cunnard are at Reading. Give them my remembrance, etc. We have not forgotten them nor in fact any our old friends. By the bye, I protest against the girls laughing about such serious matters as our cooking. Just think, our lives to say nothing of health and such trifles depends on being able to cook. Now we can cook soup, of several kinds; shin bone, potato, etc. We can cook beef steak, ham, etc., make bread, biscuits, doughnuts, gingerbread, pies, mince, apple, peach, cherry, meat fritters, fried bread, slap-jacks, mush, corn crack, and Allen is just at this present time making muffins—and that too without milk or eggs. Just tell them to beat that if they can, and if they can we are sure we can beat them eating what we cook. I really forgot to say that we can make coffee, tea and chocolate. In fact, we can do so many things in the cooking line that I really despair of being able to tell all at one time.

Remember us to Kate Althemus and everybody else that knows us. I expect Allen will write also.

Your affectionate son,
Hosea

∽ *Letter from Allen Grosh:*

April 30, 1850
San Francisco, California
Dear Father,

You no doubt will be surprised to find us still in San Francisco but so the world goes, there is no telling what a week may bring forth, especially in California. Since Hosea last wrote I have been laid up two or three weeks with the scurvy—for which—by the bye, I think I have found out a remedy—namely—Flowers of Sulfur.[15] It cured me in a week, when nothing else done me any good . . .

. . . highly satisfactory letters mailed 5th and 12 March, we received a few days ago, also the package of papers, and I am really ashamed to send so short and meager reply back. Hosea and I have the management of the household, and as we have made a grand discovery of a bed of cockles on the beach, we have been living very high—clam soup, pies, stews, fries, fritters, salad and-so-forth and so on every day, and nearly every day some of our friends as guests—for "Oyster Dinners" made out of cockles are not to be got every day even in San Francisco. My reputation as cook stands very high among our friends, and in attending to its preservation, I have neglected writing. Hosea sends the same excuse. It may be three weeks yet before we start for the mines as Farrelly has gone to San Jose to attend to some business which may take him that time. We go to the mines together. My attack of scurvy prevented us from going last month. By the bye, our gold washer, it has been laid on the shelf since my sickness, but the model is just completed and will be tried tomorrow or next day. I am more sanguine than ever. I do not see how it can fail—and if it succeeds it will be a fortune to us sure. Before going up to the mines we will send you a drawing of it.

Your letters . . . we were especially glad to hear that you can get along for a time at least without aid from us. Hosea and I will have about $100 each to go to the mines with and can if we wish draw on Farrelly for more, which I do not think probable, as $200 will furnish us with everything and fit us out comfortably.

It is almost certain that Hosea and I will be home next fall on a visit so that if John Lash has not yet started, beg of him to remain until we all then can come out together. Tell him not to attempt to come, if come he will by any other route than by Panama. He knows not what it is to come 'round the Horn—especially as a sailor—but above all things tell him not to come across the Plains. Any route but that.

Taylor and Uriah are at present living with us, and we get along quite comfortably together.

J. Winchester[16] is at present editor of the *Pacific News*—I have not yet presented your letter but will do so at an early day. We feel very thankful for the trouble you have taken about the letters from Cols. Banks[17] and English. . . . They may probably be of service to us. Judge Geary[18] is the Democratic candidate for mayor in this city and will, no doubt be elected.

I must close. But before going to the mines we will write at full length. In fact, I have a couple of unfinished letters, giving details of events, and little occurrences, which hitherto have been missed in our letters, but for want of time could not finish them. They will come by next mail.

Thanks to Uncle Moses for his kindness in procuring the letter from Colonel Banks, and am no doubt indebted to him for it.

Love to the girls and tell them we have a letter in store for them. Our health now is remarkably good.

Truly your son, E. A. Grosh

Excuse this letter—I have no time to re-write or I would have done so.

᧐᧐ *Letter from Allen Grosh:*

May 13, 1850
San Francisco, California
Dear Father,

Yours of March 25th was received this morning, through the post office. It was a drop-letter; and placed there, no doubt, by the Judge Geary, as the letter itself informs us that it was sent under cover to him, accompanied by one from Colonel E. Banks of Lewistown. Many thanks, dear father, for your kindness and attention. I shall wait on Mr. Geary at the earliest opportunity. It is certainly a great favor I receive from Mr. Banks, and as such I esteem it and hope to remember it; for Judge Geary is one of the most popular men in the city, and deservedly so, for while many of the officers of the late administration disgraced themselves by prostituting the high and ill-defined power placed in their hand to their own personal aggrandizement, he stands as high as ever in public opinion. He is especially distinguished for his integrity and uprightness of character, ... He sent for me the other day and read a portion of a letter from Uncle Moses, in which it was stated that he was going to send a small lot of boots to his care, and wished me to take charge of, and sell them. If they come by the *Sarah Sands*[19] or the regular steamer of the 20th, I will yet be here to take charge of them. If they do not, Major B. promises to send them to me in the mines. He also says a pair or two are for me probably those you speak of sending by the tin-box for Hosea and I. By the bye, we will fill the box—if it is not too large! You say that you are at a loss to know what to send as chinking.[20] We can only say you and the girls know better than we do. Everything, almost, is sold here, retail, at four or five house prices, and many little things such only as the girls can think of, would be, no doubt, of real service to us if we only knew it. To tell the plain truth, father, our wants are much more limited than they formerly were. A life in California is at best but a series of saying "if this won't do, that will," is the leading form in the faith and practice of California everyday life. What we have not, we do not want. The cook finds eggs but a poor addition—if he has not got them. Should he be short of potatoes, he crumbles bread into his hash and swears it is all the better. Up the country the miner often finds the whole of his kitchen within the narrow compass of a tin-cup—in it he boils, stews, and fries—makes tea, coffee, and chocolate—all the while declaring pots and frying pans but useless encumbrances. And thus we go—a pewter spoon and a silver fork—a plated candle stick and a tallow candle—china bowls, and a battered coffee pot, Liverpool ware[21] and tin plates, porcelain mustard-box and forks with from one to four prongs, the rarest delicacies of field and flood served up on tin dishes, are everyday occurrences at our table.

But, father, should everything be as serviceable as the box, boots and seeds, you will have made a decided hit, for all will come at the exact time, and all would be very high priced.

. . .

Hosea saw Dr. Drinon this morning, who is now practicing here. He enquired after you and expressed great pleasure on hearing that you were in Philadelphia, "Just the place for a working man," he said.

We yesterday morning received from the post office a note from Spencer B. Alden[22] of Utica, requesting us to call on him at the "American [Hotel]," as he had a letter for us from John and Emma. Hosea called immediately but he had started for the mines. He will most probably leave the letter at the Sacramento City post office as he was informed by one of the Utica boys that I was there.

Taylor and Uriah started for the southern mines last Saturday. They are well. Uriah, I am sorry to say, has suffered considerably by his store—it has left him considerably in debt, though, from what I can learn, he is in no wise to blame—Riggs, his partner, gambles, which I rather suspect is the cause of the failure though I have learned it from other sources than Uriah.

I send a couple of *Pacific News*—extra—. . . only 50 cents each!!—what a paradise for printers. . . . By the bye, I have the offer of a situation at San Jose next winter in Winchester's [printing] office. If wages keep up, I may not

come home next fall, but accept it. Hosea, though, you may count on sure—
that is, if our luck changes, or rather has changed—as I am persuaded it has,
for everything goes on right and well with us, for this past month or so. Mr.
Holland, (late of the North American Office, Philadelphia) takes charge of the
office, and him and I are very intimate. He forms one of our household, which
at present consists of two other Philadelphians, besides ourselves, viz.: C. W.
McCaulley and Isaac Lyons. In fact, I could get the situation as soon as the
Sarah Sands arrives, but we go to the mines.

I was going to—but I won't—give you an account of our cooking and housekeeping experience. I will keep that for my long promised long letter—for the
girls. Suffice it to say that we get along as harmoniously as doves, . . . we live
on the fat of the land . . . we know the haunt of the bullfrog—selling at the eating houses for $2 the half dozen—discovered the secret of finding cockles—a
delicious sort of clam, highly esteemed, and, alas! that I must say it now, just
as we are going away, Hosea returns from a rambling among the hills and
reports strawberries to be ripening.

There are five of us, with nothing else to do than to catch clams, hunt bull-
frogs, get wood, cook, eat and sleep—for Holland is waiting for his office to
come in the *Sarah Sands,* McCaulley for his father to come out in the next
steamer, and Lyons for the building of a steamboat, to be launched in a week
or two, in which he goes as steward.[23] Here we are five of as happy dogs as
you should wish to see.[24] I really look forward to the time of breaking up
housekeeping with regret.

You will see by the *News* that Winchester has the appointment of State
Printer. I have not yet called on him.

John H. Gihon[25] is elected one of our assessors—shame to the city! He is in
the gambling halls, on the square and is a quite devoted worshiper of the god-
dess Fortuna, judging from his "offering" on the Mountain—or "Monte."

Of the fire I shall say nothing this time—referring you to the paper. We
were on the ground early and never did I see anything spread so rapidly as it
did. It was a sight beyond description—never did I even dream of anything so
beautiful and yet so awful! Twice now, has the heart of the city burnt out—and
both times—almost a miracle in this airy corner of the world—there was no
wind. There is no one in San Francisco but what firmly believe that the whole
city must sooner or later be devoured by fire—a second Sodom,[26] and verily I
believe the wickedest city in the world—and yet everybody seems as careless
about the matter as though five or six millions of property is not worthy a second thought. How it has escaped so long is a wonder to every one—yet it has.

One word, the land is covered with a carpet . . . of flowers! every color,
every size! The first of May I spent gathering flowers, and visions of Rockdale,[27] sugar cakes and sandwiches kept mingling themselves up so with the

flowers that sometimes I thought I heard you in the bushes around me. But you wasn't there, though. California is a complete flower garden.

We are now watching for the seeds which we promised mother. For her we will gather them—for her you must sear[28] them, and when you cast the black seed in the ground, hope, and a vision of beauty will bless your faith. Mother! mother! oh could we but have seen you once more!

Adieu—
E. A. Grosh

N.B.—

If John Lash has not yet started tell him to wait—we are often uneasy fearing that he has shipped via Cape Horn or started across the plains.

Remember us to all—

Truly and Affectionately

E. A. G.

ℐ *Letter from Allen Grosh:*

May 30, 1850

San Francisco, California

Dear Father,

You will perceive by the *Pacific News* which we send with this that the foreigners in the southern mines have caused some disturbances by their refusal to pay the Alien Tax levied by a late act of our legislature, and also the bloody massacre of the poor Indians of Clear Lake.[29]

As to the Alien Tax I doubt very much both its expediency and constitutionality. But the foreigners owe it entirely to themselves. Their arrogance and insolence, especially those from Mexico and Chile, has all along been almost insupportable. The picture given in T. Butler King's *Report*[30] about the state of those mines, is so far as I can learn. . . . It will be something for her enemies at home to harp upon. Until this act was passed, she had done nothing but what she had a clear right to do. Necessity compelled her to take her stand as an independent state, and as she reaped none of the advantages occurring from the federal union, she, of course, claimed all the benefits arising from her independent position until that position was altered by her admission into the confederacy. She accordingly levied taxes, established post offices, exercised admiralty jurisdiction, and denied to the U.S. the right to collect revenue within her borders.

But, here, I am afraid she has clearly overstepped the mark, even if she have a right to legislate so strongly against any portion of her population. The gold mines are all situated on public lands, and those lands were purchased by all the states. California has no right to expect any other advantage from these mines, than the increased trade, prosperity, etc., attendant on their working, and the first immediate use of the gold by her citizens. But in this tax she assumes the prerogative of the general government[31] to collect revenue from the public lands for her own benefit. Supposing she had not levied this tax? It is rather hard to see the country drained of millions of dollars by those who give no equivalent in return except for what they spend for their living during their short stay here. Had the Government given us only a mint, and prohibited the exportation of [gold] dust, laying a just tax in the shape of a percentage, all this trouble would have been avoided.

The character of our foreign emigrants is mixed and . . . for the hordes of Chileans, Mexicans, and convicts from Sydney there is but little sympathy.[32] Our chain gangs—damning stain on our character as Americans as it is—is filled almost entirely by foreigners from those three countries. The greatest insult you can offer to any man is to ask "if he is from Sydney?" The people of these three countries have caused nearly all the trouble which our young state has yet met with at home. The great proportion of crime from theft to assassination can be laid at their doors.

With the exception of the "sporting gentleman"—and even they are of the better class—the emigration from the States has been of such a character as an American can well be proud of. "Public Opinion" the basis of "government" in the mines, is, I verily believe, as sound and pure, and more powerful than anywhere else in the world. The alcaldes of the late provisional governments were in almost every case men distinguished at home for propriety, honor and integrity. In many instances they were judges of high standing in the States. In many of the diggings gambling—the present curse of the country—was prohibited under high penalties—and those penalties were as rigidly enforced in the placers of California as they would have been in Pennsylvania. Crime was punished severely—perhaps too severely, even under the circumstances—and what was much better, punishment was rendered as certain as possible. The necessity of power in the government was recognized on all hands and consequently I presume there never was a more absolute, or firmly supported set of rulers than were the alcaldes of the past year. Every person within his jurisdiction became at once . . . his colleague, and his constable. . . . I would sooner trust my rights or property, or even life to a California jury than to any other tribunal in the world. Common sense became the common law of the country, and no quirk or quibble of the lawyer, either injured or benefitted you. Naked justice you might expect at their hands, and nothing more.

It is surprising as well as gratifying to see with what ready ease the Americans, wherever they may be, provide regulations and laws for the circumstances under which they are placed. Justice, law and order are held as dear by the people of the U. S. as are their institutions of republicism. Webster was right when he said that the people of California, taught the rudiments of self-government in the town meetings at home, were better qualified for their new destiny than the most enlightened men picked out from all Europe.

Since last year, however, and I am sorry to say it, the emigration from the States has been adulterated so, what in its quality by the importation of New York and Philadelphia rowdies, etc., and they, too, have done considerable towards disturbing the good order of the state. Their life of dissipation and excess, will, in many cases, be closed by this summer and next winter, for the diseases of the country will not bear trifling with.

What California now needs, is a resident population, and from every appearance she will soon have it. Our merchants are sending to the states for their families, and our working men are squatting, clearing and building preparatory to the same step. We will then have men who will take an interest in the state, and this is the one thing needful. This will sweep away gambling, for there are very few men, though they even may play themselves, who like the idea of raising a family amid the eternal chinking of the Monte table. In many parts of the mines the gambler is already excluded, and this as I said before is the great curse of the country.

It will not surprise you father to hear though after full and careful thought I have resolved to cast my lot with California this time forth . . . in my own mind she must be and will always be one of the best states not only of the Union but of the world. The surpassed fertility of her arable lands—and they are far more extensive than Butler King dreamt of—and the apparently inexhaustible mines of gold, silver, copper, and quicksilver, must, with the energetic people now gathering from all quarters of the world on her soil, be one of the greatest nations of modern times. The Pacific Ocean is her Empire, and on that field she will be supreme. Her immense wealth will enable her to gather from all quarters of the world the greatest luxuries, and the rarest works of art, while it will draw to her the whole trade of the East. Already has this begun. Look at our commerce. The Sandwich Islands, three years ago but a small speck on the face of the world, have already grown in importance to such a size as to attract the attention of the whole world. Our trade with China, before the summer is out, will rival that of New York and Boston put together, and it will not be long before she breathes the breath of new life into the nostrils of long forgotten greatness, and people the plains of Asia once more with a great people.

(Second sheet)

There's one sheet [of my letter] on politics—it is time to go to something else. I speak only for myself as to my intention of making California my future home, though Hosea, from what he has said, I should judge entertains the same views.

What say you, father, to joining your sons out here? You and Pinckney, Frank, and perhaps Bayard? Why not? The country is new, and offers as wide a field for enterprise as any in the world. We want just such as you out here, to aid in building up a great nation and establishing the family on as strong a footing as could be wished. Pinckney, Bayard, and Frank all have families; if they want to establish them as they cannot hope to do in Pennsylvania, let them come out here. But grandfather—what will we do with him, it will not do for all of us to desert him in his old age. No, God forbid. But why cannot you coax him along? Get him out here and we will make a governor of him! There, now, do you want a clearer track?

But seriously I hope for success this summer. . . . Yes, on every hand things look brighter. A few months ago we were broken in health and pocket; now we start, and the same day this leaves for you, for the mines, with enough to take us there, and with health so good that at least on my part I can scarcely realize. We were out on a hunt of a week—a short time ago—and we both returned stronger and healthier than we started, in spite of the hard walking, heavy guns, and heavier packs on our backs. Many of my friends did not recognize me, I had improved so much.

At San Blas, I prophesied to Hosea all our misfortunes. Why should we not now believe in the presentiment which promises success as well as in that which told us of trouble and suffering, and hung over me like a pall of darkness, until the news of mother's death struck down struggling hope and health and energy at one blow—the darkest hour had come! And dark indeed it was! But the crisis was past. One week after this sad news reached me, and while I was yet bending beneath the heavy hand of disease, a new light cheered me— the cloud passed away—for I felt that hereafter we would be successful in our endeavors to raise the family above the narrow, limited circumstances which we know full well had caused you and mother many a day of uneasiness and many a night of suffering. God give us strength and the memory of the past will give us heart. Sweet is labor, toil and trouble in such a cause, for such an end.

But I wonder, father, we expect to be successful this summer and it is our hope that when we have the ability to provide a comfortable home for you, the girls and Warren, that you too, will turn your eyes towards this land of promise. There are in it many bright and beautiful spots worthy of the name of Eden. Should we secure one of these and surround it with all the comforts of the country—with book, gardens, fields and orchards—would you leave the green hills of your native state, and establish your family in California? But it shall be with you to say; if you think you can live happier there than here, God forbid that you come.

I love California—every day more and more and am far more contented here than I ever was at home. Hosea, too, seems as warmly attached to this country as I do. Like many others we came here, in the last months of the dry season, when the hills were all brown and the shrubs all leafless, and we thought San Francisco [was] California. Since then we have discovered that five miles walk would have brought us to another country, where nature clothes her fields in oats, and clover in which a horse might hide. . . .

The country would suit Frank capitally. I do not know the reason, but I never think of him without thinking how well either he would be suited to the country or the country suited to him—perhaps both. After three weeks residence he would be as much at home here as he would be in Lancaster County. Such, too, are the men we want. Pinckney, what an opening would there be for him. He could garden here to his heart's content, and with some profit, for potatoes are still from $20 to $30 per hundredweight and everything else in proportion. He could find a "gold mine" or at least a "mint" without going far from San Francisco, and have an eternal spring, year in and year out, for five miles down the bay and you are out of the influence of the "Northers" of San Francisco. Our hunt, before spoken of, took us down the coast among the coast range of mountains, and instead of finding a barren waste as we were told to expect, we found hills and valleys completely covered with wild oats and clover in which a grizzly bear might hide with safety. It frequently came up to our armpits. We went down about 40 miles and returned by the Bay here. Of the San Jose Valley it is needless to speak. You can tell the girls that for five nights out of the six that we were out, we enjoyed each night a most melodious serenade from coyotes[33] and lions. Two nights we had a lion prowling about us all night. He didn't hurt us, however. The largest grizzly track we came across would have measured at least 7 inches across. Were it not for the claws it might have been mistaken for a young elephant track. The rascal frightened away a deer we were trying to get a shot at. He had apparently run at our approach or perhaps we would have went through the same maneuver. Bayard, shall we hope for him? Yes for he will find the lumber business quite as profitable here as it is on the banks of the Susquehanna. As for grandfather, as I said before, we will have to elect him governor! But adieu to dreamland!

Last Sunday evening we had an alarm of fire and Hosea and I repaired to the scene of conflagration. The fire was happily got under before it done much damage. . . .

Before coming home we stopped in at the "Orleans" a new gambling hall, just opened, and probably surpassing anything of the kind in the whole U.S. for the magnificent and costly decorations of the interior. We met Wilkinson of Utica, who after talking a little time, asked us if we "know who that was?" pointing to [a] stout, robust looking young man standing by a twenty-one table watching the game. We did not know him but who do you think it was? Ab. Hallock!—Who will we not see out here next? We found also Tomas McElvane, and William Marsh. Frank and James Williams sons of W. W. Williams, the school teacher, are also here though we did not see them. Frank is one of the proprietors of the "Orleans."

You must not think anything disreputable of Ab. for being found by the side of a vingt-et-un table[34] for the Hells of Portsmouth square are frequented by nearly every one, and in the absence of society will continue to, until some place of resort is provided of a less objectionable character. There is not a ready room in the city, nor any other place of resort, except the richly furnished, and gaudily gilded temples of Fortune, where you can always get rid of your surplus change much more readily than you made it.

Well, Ab. accompanied us home, and we almost "made a night of it" (for it must have been in the neighborhood of 2 a.m. when he left) talking over old times recounting adventures and exchanging histories. Ab. is a ship carpenter. He shipped on board the steam propeller *Endora* which, you will remember was wrecked a few days out. He got two months pay—$80—in advance—and after her wreck returned to New York that amount better. He worked awhile longer and soon made enough to come by the way of Panama, took passage, arrived at the Isthmus, got an opportunity of work [for] his passage up on the *Isthmus*,[35] sold his ticket, and arrived here with—what is not commonly the case with young men—money in his pocket. He looks uncommonly healthy, and the same old Absolum. He sends his love to you all.

John writes from Utica. . . . If he is a good fancy baker, he will, if he establishes himself in that line, do well, for we have very few of the kind in town. Cakes, of all descriptions, crackers, hot rolls, biscuit, rusk,[36] etc., etc., could hardly fail to succeed here. But I think we will be able to pick a good place in the mines, where he can make enough to establish himself here next fall handsomely. Hosea is now writing to John. We received the letter he sent out by Mr. Alden as I expected from Sacramento City.

We called on Mr. Winchester, sometime ago and he received us very kindly indeed, offering me a situation in his office, treated us to a glass of soda water, and requested us to call often and at any time to read the papers, etc. We had a long chat together in which he gave a short account of his history as connected with the *Pacific News*. He came down from the mines last February, and had not enough to buy himself a pair of boots—took the appointment of assistant editor—a month more, he was sole editor—[another] month he owned ⅔ of the paper, and in less than another month received from Governor Burnett the appointment of State Printer! He pays out for help alone over $1500 per week! He has now on hand $100,000 worth of State printing!! And says that at the end of two years, he will be worth, at the least calculation, $100,000— this he is sure of!! What an El Dorado for printers! He has sent for his family and is going to make California his home. Next fall he removes to San Jose. To show that his calculations are within the bounds of reason I would merely state that three firms before J. Winchester have already had the *News* in hand within the last year, and all gone home with their "piles"!!! Do you not feel your fingers tingle?

John H. Gihon has been elected one of the assessors of the city of San Francisco. . . . He remarked that in less than one year he would make by one speculation which he had in view his "pile" of $100,000, when he would again return to the States. I very seldom visit Portsmouth Square, but when I do I generally find him there—I am informed that he plays very heavy. We have a beautiful crowd in our municipal authorities! A sweet clique! The alderman voting themselves $6,000 per anum, each. To be sure, they are kept very busy just now, and will be for a month or two to come—after that their duties will not be a whit more arduous then than they are in any other city of the Union. But I am getting into politics again.

I met Mayor Geary one day at Winchester's and was there introduced to him by Mr. Winchester. I made mention to him of Colonel Banks' letter, which he had received. He spoke very highly of Captain Loeser[37] and said that he was one of the best officers in his regiment—that his only fault was his drinking. The tin box had not then arrived. It probably is on the *Sarah Sands,* and she will be in this week. You will see by the *News* the cause of her delay.

We received your package of papers, and several documents from Mr. Moore by the last mail. It is surprising to us here in California that Congress should make such a fuss about us out here, for doing what we poor Californians never thought of but as right. The South might just as well attempt to establish her "peculiar institution" in the moon, robbing the world of the bright light of that luminary as to force it on the people of this state. If they cut us in two they will have two free states to admit, and if they like that better than the way we have arranged things let them "go ahead." Seward is right in saying "California must be taken in now or never."[38] She cannot maintain her present anomalous condition much longer. If they wish to cast us off let them do it soon. . . . California, cherishes as deep a feeling for the Union as does any state of that composition. But why talk of the matter? I do not believe there is a soul in the whole state that for a moment thinks the Union in peril. Southerners as well as Northerners unite in declaring that chattel slavery shall never pollute the gold-strewn soil of California. Gwin[39] gives the reason—labor is honorable here.

What do you think of our stone for the Washington Monument? Is not the idea a good one? Will future ages believe that there were men in the U.S. who struggled hard to prevent the placing of that stone there? with its rich and glittering burden. This is the gift of the youngest. Altogether it reminds me of the fairytales of my younger days, in which the abused youngest always gained the most sympathy, and "came out" the best.[40]

John [Lash] in his letter talks of coming out here next spring. I think that will be the best time for him, as the city must be rebuilt of brick—even San Francisco cannot afford to throw away $5,000,000 or $6,000,000 every few months. By that time building material will I think be abundant. On the closing of the present mining season, the opening of brickyards and lime kilns will at once present an open field for the hardy miners. They will collect the capital for themselves during the summer in the mines. Last winter the project was a popular one among the laboring hardworking men, and I know several who would have embarked in the business had they had the capital. Perhaps by that time, too, we may be able to help him some. He must bring with him all the capital he can raise, for there is no place in [the] world where it yields so profitably—at present the rate of interest is from 10 to 20 percent—extra good security might probably, obtain it for less. You may judge of its value by its demand.

I should not only like him to come out, but come out to settle. No young man, with a little means, can miss it in coming to California with his family. In addition to this a young married man can obtain assistance—or in fact a married man, old or young—can obtain assistance where a single man might apply in vain. . . . Hosea, at any rate, and if we do well, both of us, will be at home next fall, and talk over matters with him. Look out or we will make Californians of you all!

Our happy little household will in a few days more be broken up. Three of us are going to the mines and the other two will soon be called to different parts of the state. I announce this with regret and sorrow for I do not believe you could pick out five persons who work together so harmoniously. However, you may tell the girls that we have made great progress in the art and mystery of cooking, for Lyons is a young sailor, and is qualified for any post on board of a ship, from captain down to cook and forecastler. He was for several years steward on board one of the New Orleans and Havana lines of packets and came out here as mate of the bark *Adelaide* of Philadelphia. He is one of the best cooks in the country, and we have been his faithful pupils. Look out girls or when we get home, we will take charge of the kitchen, and teach you the way we live in California! We go to the northern mines. The high waters have prevented much being done yet, but before the close of the month the dry season will have fairly commenced, affording an opportunity of working to advantage on the bars. I wish we had a good sub-marine armor, for I feel satisfied that this would prove a profitable manner of washing. The Yuba is a wild, rapid mountain stream abounding in eddies, and in these eddies the holes are often 20, 30, 40 ft and upwards. The only parts worked are the bars which during the rainy season only are submerged. Necessity is the mother of invention. We will find out something. Depend upon it we are going to work in earnest to make up for lost time.

As to our machine, which, no doubt you begin to think has been quite forgotten, you shall hear from it when we know what it will do. We are as confident of success as ever.

We regret very much that we are not able to go to the Trinity [River][41]—but the state of our funds will not admit of it.

One word concerning T. B. King's Report. As a whole, it is a very fair report, probably as correct as could be made at the time he was in this country. . . . I have been informed that one person—an old Californian—in San Jose Valley owns 150,000 head. Several others are reported to own near that. I doubt very much if King thought at all of the southern part of the state, when he set down that amount. 5,000,000 would probably be nearer the mark, so that it hardly will be necessary to import cattle from the Western states "by tens of thousands" for some years to come. It seems to me that almost every writer estimates the arable land of California entirely too low. Except in the gold regions one crop can be produced almost anywhere, and where the farmer has the means of irrigation, most crops can be produced _twice_ in a year, as in Mexico. But I shall know more about this some of these days.

This is a long letter, and I had intended to send more along with it for the girls—but I was unwell a day or two ago when I commenced writing and finishing up.

Direct as you have heretofore. By calling at the _Pacific News_ office, anyone can keep track of me, as we will leave word there as to our whereabouts. Do not mind paying postage on your letters, as we can do so better than you can. Love to the girls and Warren and all the rest. Please pay the postage on Marion's letter.

Truly and affectionately your son,
Allen

1851

Letter from Allen Grosh:

January 5, 1851

San Francisco

Dear Father,

I came down from the mines yesterday morning, for the purpose of taking our little property into the mountains, where we will establish our home until our "piles" are made. This is the first time either of us have been down, and you may judge what pleasure your letter of September 10th gave me for we have not heard one word from home since we left this city last June. It has relieved me, and will Hosea, of many anxieties.

Enclosed we send you a small New Year's gift, a draft on Thompson and Hitchcock, New York, for $50—It is, we know, but a poor apology for the disappointment we have occasioned you, in intimating that, most probably, by this time one or both of us would be at home again. But for either of us to have left this fall would have been ruinous to our prospects. We regret extremely that we cannot send a larger sum than this—But patience, father! It will come some day.

We left San Francisco last June for Coloma favored by General Winchester and brother with letters to some of their friends in that vicinity. At Sacramento City we fell in with two or three of our company, and a Philadelphia friend T. W. Cheesman who were preparing for a prospecting trip up to Gold Lake.[1] We had an agreement with Cheesman to mine together during the summer, and he was also to furnish funds to build our machine. A short conversation with him, and I soon discovered that, like most of the miners of that day he was inclined to believe that nothing could be made to compete with the Virginia rocker.[2] He had just returned from the Deer Creek mines,[3] and presented (though unintentionally) an extremely exaggerated picture of the difficulties we would have to encounter. Neither of us had been to the mines. It might be as he said. We dropped it. He urged us very hard to accompany him to Gold Lake, and offered an outfit of . . . provisions, but as we had no faith in Gold Lake so he went his way and we ours. We arrived in Coloma with three dollars in our pockets, and obtaining credit for a rocker and mining tools set to work, and made $3 or $4 per day. You cannot imagine our vexation and chagrin when for the first time we visited the bar on which we first worked—all Cheesman's obstacles vanished—our machine, with a few trifling alterations would have succeeded to a charm—it could not have failed! I cried for vexation. Hosea and I made a resolve, come what might, to dig from the earth the funds to lay the foundation of our fortune, and neither receive nor solicit aid from any person.

Not being able to do anything on the bar we moved higher up the river, struck a rich lead, and for a while we done very well. We worked out our hole, and were invited by a damming company to join with them, did so.

The dam "spoiled" us! About 2 months valuable time and all we had previously made were given for nothing.

Before we went into the damming operation we were joined by a gentleman from Vermont named B. S. E. Williams. We dissolved partnership only a short time ago. We differed as to future operations, and parted, as we had lived together, the best of friends, and with no hard word or feeling, and a perfect understanding.

After the failure of the dam, we moved some two miles further up the river, struck a small lead and in about a week, cleared ourselves of the store, and purchased a fine jackass for four ounces. He is a powerful animal, capable of packing 375 pounds. . . . We struck off into the country . . . We settled down about 7 miles to the south of river on a little creek called Indian Creek, 11 or 12 miles southeast of Coloma. Here we opened a hole which yielded from 25 to 50 cents a bucket. We had to pack our dirt about ¼ of a mile. We purchased a jenny, and were doing a first rate business, when Hosea became unwell, and in a day or two Williams was down with one of our peculiar fevers. I was hardly able to work myself, and for a month we done little else than to spend what we had made. We all got on our feet at last. By this time the diggings had been pretty well worked out, and receiving information from a friend of a new region, off we started again. We left camp during the night to prevent our destination from being discovered, and it was well that we did, for the country, we afterwards ascertained, was scoured for fifty miles around in every direction, and had our locality been discovered, we would have been so crowded by our old neighbors as to render a profitable winter residence there impossible. We settled down about 2 miles north of the Cosumnes River.

Our healths by this time were pretty well restored and for about 3 weeks we worked like devils, throwing up dirt.

Our dirt disappointed us, but still we cannot complain, for we got about $200 apiece, besides living, etc.

This is sufficient for a good start. I am now here for the purpose of buying materials for building our long talked of machine, which is nothing more nor less than drawing through a bath of quicksilver, by means of a siphon, a continuous stream of dirt and water. We depend not so much on the amalgamation of the gold as we do on its specific gravity. While the gold sinks all other substances with which it is mixed rise to the top. By the shape of the bath an undercurrent is created in the quicksilver, which draws the gold which has a tendency to sink, towards the bottom, out of the influence of the current which discharges the dirt and water.

The screens, riddles, etc. are rather complicated in description. When once in operation I will speak further of it.

Even should this machine fail which is almost impossible—but even should it—I have within a few weeks past effected an improvement in the common rocker which will increase its capacity ten-fold, and place it on equality with the quicksilver rocker in the economy of its working.

Thus closes our first campaign in the mines, and if we have not made our "piles," we are good miners, and have sufficient experience to carry us through another year with credit and honor. It is true, dear father, that had we been so minded, instead of counting our little pile by hundreds, we might have thrown it into thousands. A few days before leaving San Francisco last June, Hosea was offered a situation as clerk at $300 per month and I one in the *Pacific News* office in which I could have made from $75 to 100 per week. We refused them and I think the result will prove our wisdom.

We have learned the business of mining and for this object we have sacrificed everything else. We have left no toil, no fatigue no pains nor expense between us and our aim. We know the whole county[4] in which we intend to operate. We know the location and directions of the richest quartz veins in the region, and already have possession of small but very rich one.[5] This knowledge alone should enable us to acquire a princely fortune and perhaps <u>will</u>.

We have, thanks to your kindness and that of our friends in Pennsylvania, whenever we wish to avail ourselves of it, an entrance to the best society in the state, and can already look forward to an advantageous settlement in any portion of the Pacific Coast. Our long (do not smile father—<u>comparatively</u>) residence in California go where we will, is already of great advantage to us, and the original Californian wherever we find him, hails us as friend. But enough of ourselves.

San Francisco is indeed a wonderful place. Compare the two views in Bayard Taylor's *El Dorado* with that accompanying this.[6] It has advanced [in size] since those days, nearly ½ a mile or so into the Bay. On landing I struck a new street and walked some fifteen minutes before I could find out where I was. Sacramento, too, I found had out-grown my recollection. "Wonders sure will never cease."

Farrelly my old chum, has settled at San Jose Mission, and is doing first rate. N. Witman is in Monterey, and doing well. The rest of our company I know nothing of.

H. Gihon has gone home with his pile. He was tolerably popular here for a while but times are changed, public opinion became frail and he found it time to slope.[7]

Please thank Brother Thomas for me for the hint he gave me concerning Winchester.[8] It has not only benefited me but also a warm true friend to whom I gave the wink. Winchester holds all his property in his brother's name, and at the same time he is his brother's trustee! He is as slippery as an eel.

The report of the freshet on the Schuylkill [River][9] reached us in the mines, through the means of a stray Boston paper. It was indeed a melancholy affair, and I fear Reading has received a blow which will put her back some years.

I congratulate Emma and John on their fortune. Assure them that on my return here (about the middle of next month) I shall take care to redeem an old promise. We received a letter from Mr. Spencer[10] last summer from Sacramento City. We answered it very unsatisfactorily, for we were then on the que vive[11] and knew not where the next day might find us. His letter probably laid so long in the Coloma post office as to render the answer unavailing.

I am grieved to hear of Aunt Mary's ill health, and hope it may not turn out as Frank fears.

Tell the girls that I forget the Mexican chocolate receipt[12] but will get it from Hosea.

The biggest lump [of gold] we have yet found we got last Saturday week. It weighs $38[13]—If we can keep it we will send it home some of these days.

But I must close before any letter is half written. I have at least half a dozen more yet to write, which cannot be put off. Oh for one short half day's conversation with you all.

With this, rest assured. The future never before promised so rich a reward for our labors, and we never before so panted for action. Health, youth, strong arms, and stout hearts, and two years experience. What more could we ask? We are in California!

Love to all affectionately your son,
E. A. Grosh

✐ Letter from Allen Grosh:

January 12, 1851
Sacramento City, California
Dear Father,

Being unexpectedly detained here over today I cannot but give you some revelations which were undeveloped when I finished the letter I send with this. Hosea and I are, in all probability, rich men! During the summer past we have explored the country lying between the South Fork of the American and the North Fork of the Cosumnes rivers pretty thoroughly and from the information given me below, I think we are knowing to two or three and have possession of one, extremely rich veins of gold bearing quartz. I mentioned incidentally to my friend Holland of my being acquainted with several veins in our region, and gave him descriptions of several specimens we had found. By him I was at once introduced to Professors Nooney and Shepherd[14] both geologists to high standing here, and both engaged in quartz operations. Of course I was obliged to conceal locality, etc., but still, could give sufficient information as to the eligibility of the situations, and some few facts illustrating their richness, etc., which seemed to create some little sensation between the two professors. The vein we have possession of Hosea is now exploring. We have had our eye on it for this couple of months past, though from the information Professor Nooney gave me it must be rich twenty-fold beyond our highest calculation. It is partly decomposed, and by breaking up the rock with a pick, and washing with common care the clay and fragments of rock in a pan 15, 20, 25, 30 cents have been obtained to the pan. The professors were very kind, and when they perceived that I was not disposed to give information as to whereabouts, did not press me on that point, but gave me a full and clear exposition of the manner in which they had severally pursued in the formation of the companies in which they are at present engaged.

The vein, though small, as I said before is extremely rich at the point at which we have struck it. Its extent we do not know, though guess extends for ½ or ¾ of a mile in length. On the Carson (Professor Nooney) and the Mariposa (Professor Shepherd) they claim 210 feet to the man, with an additional claim of the same length to the first discoverers. Holland at once entered into the spirit of the thing and hinting the facts to a few monied friends, we soon found that there would be no difficulty in raising any amount of funds by sale of stock we might want. But this is not our course. Holland, Farrelly, Hosea, myself, and three persons with whom we have mined during the last summer, whose honor, industry and fidelity are undoubted, together with three or four others who though never mentioned to you before, have stood the test of a California friendship, will immediately organize, take possession, and work the mine. The vein from its decayed state will work extremely easy, and the simplest cheapest machinery will suffice. The outlay will be trifling. The capital invested will be, principally, our labor. From this mine we expect the means for commencing the others. The company, as soon as it has the means, will, under a new title, or change in the name of the firm, open the other veins, put on the works sufficient to pay the expenses of organization and operation, and next spring—and there is going to be a great rush into this business—then enlarge the company, increase the stock, and either get in capitalists to assist in their working or sell out for cash down, which last will no doubt be the safest, and perhaps the most profitable way of making money. Our best mine we will hold on to. Mining henceforth is our business.

Everything is in Hosea and my hands. Our right as discoverers, will have its due weight in the organization of the company. So do not make yourself uneasy as to us being overreached, etc. The company will be composed of tried friends, besides 2 years experience in California has made considerable of businessmen of us.

I give no particulars of the riches of our mines, not having time. Suffice it to say that I have examined a large number of them from the most celebrated mines in the state and ours will suffer nothing from the comparison. Our large lump of gold I mention in my other letter ($38.00) is from the vein we are going to commence operations on. Professor Nooney pronounced it a rare and valuable specimen—such as he had never before seen. It is extremely porous.

I will get off tomorrow morning, probably—certainly in the afternoon. I come down again about this time next month. Till then <u>adios</u>. Love to all.

In haste truly your son,
E. A. Grosh.

P.S. Please use this letter carefully. We do not wish the story to get out, etc. The injunction, of course, does not extend to the family and friends.

∞ *Based on handwriting it appears this letter is from Allen Grosh:*

April 26, 1851
Mines[15]
Dear Father,

Messrs Colson[16] and Dennis, of Gardiner, Maine with whom we have been associated for the past six month go home today, and we take advantage to send you a few lines.

They go rather before we expected and are therefore unable to send any [gold] dust as we should have like[d], having lost some little, by having a valuable horse stolen from us, and in fruitless attempts to recover him. The rest of our funds are invested, little as they are, so that we cannot well withdraw them without loss. In the course of a few month[s] we expect to be able to send at least a little.

Our prospects are at present very flattering, in every way, and we expect during the coming year to make up for past misfortunes. Our machine is not, and will not be for some weeks yet, in operation, for various reasons. The most important of which is that by the delay of a short time, we will have a large and very rich region all to ourselves. We are now and have been for some time making ourselves acquainted with the peculiarities of the country, so that we probably are better informed on it than any two other persons in California. In a short time the water will all be dried up and this section deserted when by throwing a little expense into the fitting up of the various spring[s], we will be enabled to do a large business, as we will have an ample supply of water, and a clear field. This section, the Sugar Loaf Mines, is very peculiar and very few miners have discovered the secret of working them to advantage.

I have only time to say a few words and must hurry on. The postal arrangements of this state are now so far perfected, that we can trust our letters into the interior. Hereafter address us at Coloma, El Dorado County. It is but twenty miles from here, and nothing will remain in the office over two weeks, as there and back is but a short Sunday's ride for California horses.

General Winchester has played Holland a most rascally trick in the *Monterey Gazette,* the consequence of which was that he failed in his paper.[17] Winchester has played a most disreputable part before the California public and has run to the end of his rope.

We have not yet got our letters from Monterey, but will soon.

Please write to us as to the whereabouts of John Lash, and as to his willingness to come out here yet, and assure him, though we have apparently neglected him so long, he has ever lived in our thoughts, and now though late will soon be in a way to redeem our promise. Also, ascertain of Reuben if he would be willing to come out here on the assurance of $3000 per anum with a handsome prospect of more. But understand, not that we promise him that, for repeated misfortune has made us a little more cautious, but there is every human probability that in a few months we can give him at least that, and we want to know if he will be willing to come for that, and link his fortune with us.

We are in one of the finest and healthiest parts of California, and as we look upon the mountains around us and the Sacramento Valley beneath us through the bright bloom of spring, we are forced to exclaim, what land is like it. In heart, in soul, in spirit, we are Californians, and nothing could tempt us to exchange our lot as miners, at least while we are poor. O father! You cannot conceive the wild pleasure, the deep, real, intoxicating enthusiasm that creeps into the heart, and thrills the nerves, as driving at full speed over the mountains on a good horse your eye takes in a range of hundreds of miles—on one hand eternal snows—on the other perpetual bloom.

Our partner Mr. Williams[18] wishes us to request you, if you can, to give the name of the Universalist preacher in Ohio, so celebrated in the cure of the cancer. His mother (in Vermont) is severely afflicted, and he wishes to procure his assistance.

Messrs Colson and Dennis may probably visit Philadelphia. If they ever do, please treat them as our warmest friends, and it is with pleasure and pride that we point to them as examples of our intimate associates in this country.

Enclosed we send the duplicate draft of last January. Love to the girls, Warren, and all the folks.

Ever your sons,
Allen and Hosea

✍ *Letter from Allen Grosh:*

May 10, 1851
Mount Sugar Loaf Mines, California
Dear Father,

We have just heard, this evening, of the fires in San Francisco, Stockton, Benicia, etc., and thought it best to advise you that we are in no way sufferers. We only have the news from rumor, and I hope much exaggerated, for make the losses as light as you can, it must have a very injurious effect on California.

We wrote you a short time ago by the hands of our friends and former associates, Messrs Colson and Dennis of Gardiner, Maine, and as it may be some time before it reaches you, we will take the liberty of repeating a few items:

Our prospects are very flattering, and we will settle down where we are at present located—at least for the summer. We have and are now gathering round us the comforts of a home, and the present year will pass far more pleasantly than did the past. We yet want a cow, and half-a-dozen chickens, (which in a couple of weeks we will have) to live like princes! We have an extensive "Programme" for the coming summer's campaign, and it will yet be several weeks before we get it in operation.

We request in our Colson and Dennis letter information of John Lash, and here repeat it, for you may not get it until after this comes to hand. Please inform us if he is yet anxious to join us. We are anxious that he should, and our repeated misfortunes have alone prevented us from doing what we promised him. We bought a violin for him last Christmas, and it yet hangs, stringless and tuneless over the cabin fireplace!

We also wished you to ascertain of Cousin Reuben if he would be willing come out here for a guarantee of $3000 per anum, with a good chance of making more. We are in one of the healthiest and finest parts of California and we are sure that they will like the life.

I have a severe headache this evening, and will therefore conclude. When we get things settled around us, we will give a full description of our doings, plans, and situation.

Hosea goes to Coloma tomorrow and will mail this.

Hereafter address us at Coloma, El Dorado County.

Love to the girls and all the folks.

Affectionately your son,
E. A. Grosh

Letter from Hosea Grosh:

June 7, 1851
Sugar Loaf Mines
El Dorado County, California
Dear Father,

We received your letters of February 25 to April 7 last evening and not having the run of the mails, I write this short letter, that you may not be in longer suspense than is avoidable. In a few days we will write in full. In regard to the company affairs; we twice prepared a full report of the affairs of the company; the first time in November of '49, and concluded not to send it with my resignation, because the stockholders had never confirmed my appointment as they were bound to do by the articles of agreement, nor even in any way acknowledged it; I therefore considered myself only responsible to the company as they alone appointed me. The great complaint that I have to make against the stockholders is that they never wrote to any of the officers of the company except Taylor in consequence of which we never saw any letter or communication from the stockholders which was received after our arrival at San Francisco. But notwithstanding my resolution on this subject, I prepared another report in February of '50 which was accidentally mislaid and supposed to have been sent. I have carefully preserved all vouchers, receipts and orders, kept copies of reports, etc., so that it will be a full and satisfactory (I hope) statement, when I complete it, of the affairs of the company so far as concerns myself. In regard to the others, they must look to Taylor as I know but little of them. In my next, we will send a full statement of the affairs, from beginning to end, so far as we know them with our views, etc. So I wish the stockholders to suspend judgment until that time. Make yourself no uneasiness on the subject therefore, I beg of you, as I doubt not all will be satisfied with the statement and if necessary we can produce proof and voucher for all we will say. At least a week will pass before we can carry our next to town as the country is somewhat unsettled at present and we do not like to risk riding around alone unless absolutely necessary nor lose more time than necessary so Sunday is about the only day we ever go to town as on that day we never work. We had an alarm Wednesday night. Just after the lights were all blown out, we heard a noise resembling the cry of a coyote, which we at once pronounced an imitation; a few moments our black mule snorted out an alarm; Allen looked out, saw nothing, then lay down waiting another alarm, in a few minutes the mule snorted again. (This animal is invaluable because she rarely gives a false alarm but is always sure to give one if anything comes near at night.) Allen then got up alarmed the rest, two went out to reconnoiter, in a few minutes found signs of Indians. We ran down with double barreled guns loaded with slugs, after another look around, suspecting some bushes, I fired one barrel into them, and roused an Indian in another direction; Steele, the one with us, raised his gun but it missed fire, so the fellow got off. We immediately tied the animals near the cabin and set a watch. We heard occasional signs in the distance during the night, but nothing alarming until just at daylight, the black mule snorted and pointed to some bushes some 100 or 200 yards from the cabin. Steele a brave but rather rash fellow advanced without calling the rest of us; when within fifty yards an arrow passed through a serape he wore, his coat and shirt and passed under his arm without touching him; he threw his gun to his shoulder as another grazed his arm and fired a load of slugs at the bush and rushed on; as he reached the bush he fired again with what effect we were unable to discover. This closed the affair.[19] We prepared for an attack the next night by fortifying the cabin, determined to keep it instead of going out, which however well it worked, the first time, it was dangerous in general. They came around again at night, but finding us alert kept at a respectful distance. Last night was undisturbed. Since we know our danger we are so well prepared that our risk is very slight. We at first believed it to be the Indians in the neighborhood and marched, next morning after the first attack, against a rancheria[20] near us; but on reconnoitering we found them perfectly unsuspicious of danger, chatting and laughing, we concluded that we were mistaken; so we walked into the rancheria to the astonishment of men, women, and children; but on our stating what were the reasons of our appearance they were very much frightened; after talking with them for a while, we found every moment fresh reasons for supposing them innocent.

After a while they came to our camp and assisted us in searching out the traces of our besiegers. From everything we could gather we are inclined to think it a small detachment of routed Indians from the mountains, of perhaps a dozen—certainly not more. However, we are always ready night and day, this though a little harassing will not last long as the soldiers are out and will soon scout this section or if it does, we are used to it, so broken sleep is of but little consequence. By tying the animals around the cabin we are sure of warning, especially since our black mule is a Sonoran and therefore used to Indian attacks and snorts as loud as a trumpet.[21] We are so situated that we can hold out against almost any number, we are four and thoroughly armed, having 2 shot guns of large bore, one of them double barreled, 3 rifles, 1 large Colt's revolver,[22] 1 small pistol, my carbine which is almost as good as a revolver, besides knives and such things in abundance. I have given so full a statement that you might know the exact danger which you will doubtless still exaggerate it as you do not know the cowardice and ignorance of most of our Indians and their fear of the whites; few of them have anything but bows and arrows. So our danger is not as great as you will probably apprehend notwithstanding my full and careful account. We will take good care.

Private

Owing to these troubles and some other reasons we have not been able to prospect much out of our own ravine. One of quartz vein that we thought rich did not pay enough to pay for working a couple of others we wish to look at more thoroughly. They have every indication of gold and are so situated that no one unacquainted with their situation can probably find them. They are all the safer for these troubles as few or none care to prospect in troublous times. In case we find anything worthwhile we will immediately inform you and get power of attorney to hold stock in the name of some particular friends at home to whom we will give it which we will be able to do, holding about seventy yards to the man; as we do not wish to form partnerships with many persons in the country, none unless we know them and some because we do know them; between the two classes our real friends are few and we wish none others in, as partnerships are generally closer bonds in this country than at home. We are able to do well at mining in the ordinary way and have a large field before us. We have tried several experiments with the machine none of which succeeded to our satisfaction owing to the imperfect models. We have not done more as I am loath to lose more time than necessary while we can make from $5 to $10 per day which though slow counts up surely and you recollect I have rather too much of the "Slow but Sure" to give a fair certainty for the most flattering uncertainty.

The news in regard to the returned Californians was news. We knew not of the return of Klapp or Abbott and nothing of any of the rest of the company except Green and Farrelly. Green was at San Francisco at the same time with one of our friends who describes him as particularly cross so I judge he had but little success. That is all I know of him. Farrelly is doing well with a farm in San Jose Valley, near the mission of that same name. In regard to those returned I am not much surprised at their conduct, however strange it may appear to you. From Abbot, I expected but little better. Sam Klapp has probably had but little success, and too proud to own it, and perhaps suspicious that the stockholders may be trying to gouge him. He, as well as most of the company, blamed the stockholders on account of many things, of which Taylor should bear the blame except, that [in] making him their sole medium of communication, they gave him the power.

In case of any discovery we probably will be able to have a good part of our family out here which we will be more anxious to do as we will then be able to pile up dollars and have the society of our friends at the same time. Besides we will need heads as well as hands, besides we will help to dispose of our funds to the best . . .

Hosea

∾ *Letter from Allen Grosh:*

June 8, 1851
Dear Father,

I find I must add another sheet, I leave this afternoon.

Accompanying, you will find a prospectus for a paper by B. E. Holland, late of Philadelphia. We have been since last winter the warmest and truest friends, tried as this country alone can try men. He insists on my using my influence with you to become an occasional contributor and will not take a refusal. The *Gazette* will be sent you regularly and without cost. If you would, father, occasionally send an article or a letter it would please me, and I therefore join in his solicitation. The "*Gaz*" when once started will be as strongly backed as any journal in the country.

I cannot find that tin box. It is probably lost.

Winchester has started a handsome daily in San Jose called the *Argus*, in honor probably of the Albany [newspaper], as it is democratic.[23]

The Governor's Message, which you will receive with this will strike you as rather singular, but it is necessary to measure it by California and not Pennsylvania. It will be hard for you to believe the governor to be opposed [to] capital punishment—yesterday I was so assured by an old friend of his. To be candid, though I am a strong advocate for the abolition of the gallows, I fully endorse his recommendation on that point—If we catch a horse thief in our diggings, I do not think he will leave them again.[24]

Do not think me harsh or bloody minded father. We are so accustomed to take care of ourselves that we would be ashamed to ask protection from any quarter. For two months now, I have not laid down without looking at the caps on our arms.[25] Two of our party have already been shot at by Indians (fortunately in both cases without injury) and for weeks together we have went to our work armed. We are always ready, and I think you have nothing to fear for us,

Affectionately your son,
Allen

Hereafter address us at <u>Monterey</u>

ᥫ *Letter from Allen Grosh; because of context, this undated letter is placed here, having apparently been written in the late spring of 1851. A notation in a different handwriting has been made on the letter that "This letter was opened by Malvina and Emma. We are all well and Malvina returned from Stutgner today, JGJ"*

Saturday Night
Rose Spring
Dear Father,

We have had a busy week of it prospecting quartz veins, and I will only send you a few words.

We are opening <u>four</u> veins; in two we have found gold, though in small quantities—all four are exceedingly promising in appearance. We opened a hole this evening, and though we did not "raise the color," its appearance is such as to give us the most sanguine hope. It is one mass of decomposed sulphurets[26] of iron auriferous in character, with only traces of quartz intermingled, so greatly do the pyrites predominate. This vein runs by the side of the ravine we worked out last winter and crosses it twice. We opened it at the upper crossing and found it as above described. The ravine did not pay here, but some fifteen or twenty feet below we found three lumps in pyrites, worth about $80 each, besides many smaller pieces of similar character. Our next hole we sink to strike the vein here. From fragments we judge the vein is more mingled with quartz which probably is the reason why it pays at this point and not above the matrix at the last named point not have the consistence, in the melted state, requisite to retain the gold. If this is the case we will probably some day here make a "rich strike." The rock in this vein is certainly the best looking <u>we</u> have ever seen, and has been pronounced by a competent judge the <u>best</u> in California. Some three or four hundred yards lower down the ravine the vein has the appearance of a common quartz vein and here we have struck the "lead." It is but about an inch wide, and contains but little gold. Its situation is peculiar. From a boulder swept from the vein 40 or so yards below, the lead appears to be more than a foot in thickness in the body of the vein. But at this place the vein is perfectly sound, and stands now as it was originally thrown up. It is about 5 foot thick and all the "leaves" or lay-

ers of quartz are gathered together at the top into a space of
3 inches, thus: [*see the following figure*], we therefore expect
the lead to widen and grow richer as we go down. Last week
some Indians, a few yards below this picked up a boulder of
a couple of pounds which contained $21—this vein sinks and <u>we think</u> we can
trace it by the burnt greenstone rock on the surface for about ¼ of a mile to
another vein of great promise, though we have not yet found gold. This vein
last spoken of also presents the appearance of anything but quartz containing
great quantities of sulphurets of iron and foreign matter. It <u>may</u> be that it is
<u>the</u> vein of this section, and all our others are only shoots or spurs—it [is] ten
feet and more across at the top and probably widens.

The two other veins we have but commenced on—one contains gold in the
pyrites; the other is promising.

Quartz prospecting is very hard work—it taking from 3 to 5 days to sink
a hole on a common sized vein—large veins in proportion. We only go a few
feet into the vein (from two to four) in each place unless gold is found pretty
plenty. In this way you can prospect much faster and have a better idea of
the vein than by the old mode of sinking shafts of 10 or 20 feet at hazard. All
gold bearing veins have places (and in some there are many) which contain
no gold. In the Union vein at Mathews a few miles from here, a shaft of 13
feet was sunk by some friends of ours (old and successful quartz hunters) and
they did not "get the color" on a test. Three months after, a hole was sunk
about 25 feet from this shaft, and some of the rock yielded as high as $10 per
pound. This vein (or rather this portion of it) is now held at $3,000,000. The
great Carson River[27] vein is another instance. The valuable portion of it is confined to one lead—the crossing of two veins. When I last heard of it they had
sunk their shaft 45 feet immediately on the lead, which still retains its original
shape—that of a T. $80,000 has been blown out at one blast!

You may wonder, dear father, at our opening so many veins at once. A vein
exposed at a poor place is the best means of securing it, as the initiated only
will look at it. Of these there are comparatively few—the country large—and
they never interfere with each other. Nearly every miner in the country has
"tried his hand" at the business, and give it up in disgust, either finding nothing or not knowing the value of his discoveries. We were offered a week or
two ago a fine vein of 4 or 5 claims for $2500, and could probably get it for
near that today, when I feel confident that with $500 worth of machinery and
4 horses I could take it out in a day!

I could write all night I have so much to tell you, but I have a couple of
hard weeks work before me, and must save myself for it.

Sunday Morning

We have abandoned all idea of washing more, and will turn our attention altogether to quartz mining. The present is a time which comes not in a hundred years and I think we can manage matters so that in the end we will all have what money we want.

We are very sanguine that we can build machinery that can do more work, and better with half the power, than anything in this vicinity. We think $500 and two horses sufficient to build and work a mill capable of crushing 10 ton per day. We will use four rollers for crushing and one for grinding.

We thank you heartily for your suggestion concerning the action of fire and water on quartz. The burning of the quartz is now adopted almost universally—but plunging it into water when heated is entirely new. We have made only imperfect experiments but we think it will easily make a difference of <u>one half</u> in the quantity worked besides making the powder finer. In many veins the bulk of the gold is invisible with a good [magnifying] glass, and unless the rock is reduced to an impalpable powder the waste of gold is very great. I know not a mill in the country which does not waste from ¼ to ½!

Dear father, I have given you a short account of our doing and hopes—Just so they appear to us—we <u>may</u> be mistaken in our impressions—we <u>may</u> make an utter failure of it, but <u>I see not how</u>. Depend upon it if everything fails us <u>now</u>, in three months all <u>will</u> be right. We have been for some time planning a quartz mill and we are <u>confident</u> two horses are sufficient to crush 10 ton per day—and crush it <u>completely</u>. With such a machine we can get as many veins as we want, i.e. for putting up machinery on one vein, we can get another equally good. A vein is not considered worth working if it pays less than 5 cents per pound.

Our object is to hold on to all veins we may discover worth working pick out <u>the best</u> for our family and their immediate friends. On the rest we will erect machinery and give to our friends, retaining an interest of from ¹⁄₁₀ to ¹⁄₂₀ in each.

It is not necessary for <u>all</u> our folks to come out here. A few to overlook and manage will be sufficient. But I <u>do hope</u> most of you will be induced to make California your homes. I love California and if I do well, I feel bound to do for her what I can—she needs <u>good families</u>. I sincerely believe that you and our family and their friends would do more good here than it is possible to imagine. Think of it dear father, and all the rest, <u>send the names</u> of those you wish to come out with you, and I think we can induce them to come out, if wealth can induce them. But to do this we <u>want the names</u>, so that we can attach them to quartz claims—If after the claim is taken if they do not chose to come there will be no harm done.

Now to answer some points of your last letter. Tell John and Primer to wait until they hear from us. In two weeks we will certainly write. Warren would be worth everything to us, had we him here. Reuben, too, we would have work for. But we need every cent we can raise to make a start—we will send money as soon as we can, which we trust in God will not be many weeks more. This day two weeks, by the furthest, we will write again, and sooner, should we do anything extraordinary. Remember us to all our friends in Utica. Love to the family generally. And in the hope that the day is not far distant when [we] shall all meet again.

I remain, affectionately your son,
E. A. Grosh

Direct your letters henceforth to Mud Springs, El Dorado County, California.

This post office is only 10 miles from us and one or more of us go to it every Sunday—It has just been established.

Letter from Hosea Grosh:

August 17, 1851
Dear Father,

We see by your letter of May 26th, I believe it is, that you are thoroughly wearied by the vexation of your present situation. I am sorry that we cannot as yet give you more than hopes in return. We have been working for some two or three months with but little success till within the last week when we have found a good prospect of ground diggings. We have fairly commenced opening one quartz vein which promises well, we are assured by a young man of great experience in this line, having assisted in opening near 300 veins, that the appearance indicates the certainty of gold in the vein and almost a certainty of its being rich, and permanent.

We think that there is a strong probability of this vein connecting with others in the immediate neighborhood in which case it is advisable to secure all the connecting veins to prevent the trouble that often arises from conflicting claims. We therefore send home for proxies from six persons, this will secure us from any interference whatever. We hope that before another year we can send home for the family including Pinckney and whoever else we can induce to come. This will prevent the greatest privation of this country lack of society all privations are more than balanced by the freedom from those vexations which shorten and embitter life of which you have had so large a share. I think you and the rest will be pleased with this country and am glad to see that all of the family have thought that they can give up the old states for this for I am convinced that neither Allen or myself would ever be contented in old Pennsylvania.

We calculated to start a garden in the spring, or rather beginning of winter for that is the time to commence such things here, so that we will have abundance of vegetables of every sort the next summer. We will purchase a plough shortly and use it in getting out dirt in such places as will allow of its use and by then with a scraper we will be able to perform as much as ten times the labor without working as hard. We have associated with a young man from Arkansas who was brought up on a farm and who we have known since last September and know him to be a good fellow in the better senses of the term. We are now without neighbors, the nearest miners being what is equivalent to 4 miles off. This section of country is completely deserted but think we know some good places to throw up for the ensuing winter which we rather think is going to be tolerably severe, not in cold but rain.

I caught cold last week which settled in the muscles of the back and right side and prevented my working for a couple of days past but I expect to be at it again next week. Steele is going to Coloma tomorrow therefore I write this though I have but little to say but that our general health is good and we enjoy ourselves well, feeling confident for the future and lacking nothing at present except the presence [of] our father, sisters, brother, etc.

Hosea

The proxy power that I speak of in the first page if written in the following form will answer every purpose in this country.

⮑ [*The Proxy Form*]

September 1851

I do hereby empower E. A. Grosh to act as my proxy, and as such, to cast my vote on all matters coming before the Consolidated Mining Company of Rose Spring.*

We have picked on yourself, Pinckney, Bayard, Frank, John Jones, and grandfather as persons to send proxies. If any difficulty should occur to getting any of the above their places might be filled by Letitia and Malvina, but if possible the list given will be preferable,

Yours affectionate son,
Hosea.

The following is an addition by Allen:

*Head your proxies only with the day of month and year—leave out the place. This may prevent some trouble to us, as there are persons who would

dispute their authority if they were dated out of the state. It may save us litigation—though I have no doubt it would be held valid in our courts if dated Pennsylvania or New York. We have strange laws and lawyers out here.

E. A. G.

The following additional text is on the back of the sheet and penned by Hosea:

Having nothing else to say I will put down one of our day dreams.

The plan is to get our own family and Pinckney's together with some of the rest out here and settle down in one year. We can raise everything we need in the gardening and farming keep a little stock and have it increase. In this way we will have everything necessary and keep growing from competence to wealth here in the mines. We will have market in our neighborhood for all we have to spare for many years. Besides simple gardening and raising stock we will [go] into fruit. In two years we will have strawberries etc., in three, peaches and cherries and soon until we have orchards of all kinds. I think that nearly all kinds of fruit will succeed here. If we became wealthy we would buy other ranches below and raise everything the country can produce.

All this will follow easily if once our family settle in this country even if our mining should fail of which there is no probability.

✐ *Letter from Allen Grosh:*

August 17, 1851
Rose Spring, California
Dear Father,

Hosea is unwell today and hardly fit for writing—he has in consequence omitted several things which should have been mentioned.

Quartz veins are held by claims of so many feet to each man, the discoverer entitled to a double claim—the claim varies from 100 to 300 feet, according to the section in which it is located. The rules and usages are ill defined—in fact there is very little rule about it.[28] Everybody is now (and will be until the close of the season) engaged in hunting veins, and enough are found to prevent miners from examining too closely into the validity of claims now held—"provided always" you keep the true value of your rich veins to yourself, which is no difficult matter if you can keep your tongue quiet. Excitement, like "vaulting ambition" always over leaps itself, and a vein, which properly worked, will bring you in $10,000 to $20,000 per anum is not in popular opinion, worth a thought—at least a quarrel. A man who discovers a vein generally takes it up in the name of his friends, and then associates them with him in the working of the whole. Now an idea has entered into our heads. Why can't we provide for our friends, though they may not reside in California, as well as other folks.

As I wrote you last spring, we have all along believed that some veins in this section were rich. We wished to secure them all (not as do some folks, having before our eyes the fear of being too rich) as they are conveniently situated for being worked with the same machinery, and trusting to Providence and this hidden situation left them lay until we could open them without interruption. We are now opening two, one of which promises much—the other is not opened sufficient to judge yet. We have also been hunting for a third—a large spur of the first we think which promises a great deal. A few days work more and we will probably strike the main vein among the boulders. This last Hosea has not mentioned as it probably slipped his mind. The gold from this vein occurs in lumps embedded in sulphuret of iron—which is regarded as a never failing sign of richness and permanency. The other veins we have not yet examined.

Now, father, our course is this; we have organized ourselves into a company and as we find the veins will pay for working we will take them up in the name of the company. We are entitled to four claims of 150 feet on each vein and adding you, grandfather, Uncles Pinckney, Bayard and Frank, and John Jones with Mr. Steele's father and two brothers, we can by stretching a little go close on to half a mile on each or a little farther should the vein reach so far. For a month or so this will easily be done—by that time we can show your authority for acting in your name when no one will attempt to dispute the claim.

In a short time—say 3 or 4 weeks—we will know if the veins really are what they promise to be. If they are, we will soon have up extempore machinery capable of crushing one or two tons per day, until we get sufficient to get complete machinery from the States. When we ascertain it to be a "dead open and shut" thing, we will expect you, Pinckney and John Jones to join with us, immediately to be followed by Frank within the year. Grandfather I do not expect will spare Bayard; however they must take it alternately and visit, at least their gold mine!

The veins utterly failing we count on a very profitable winter. The springs on which we counted for this summer work have failed, however, I do not think we will lose by it in the long run. We have made some discoveries which will prove valuable the coming winter. Gold mining, though a hard, is a fascinating business. Father how we do wish you were with us!

I know you will not approve of it—but we have not yet reported to the stockholders—We were induced to hold off in the hopes of being able to assure them from loss, (from sources aside from our veins). During the coming week we hope to be able to give them such an assurance, and then they will hear from us. We are not indebted to them for much courtesy, and they may wait a few days. Had they done their duty to us, their $10,000 would have been made six months after our arrival in this country. Of this rest assured—our reports—at least Hosea's, in relation to the money, are full and clear, and will vindicate any doubt upon our characters.

I write in haste. Love to all.

Truly and affectionately your son,

Allen

๑ *Letter from Allen Grosh:*

October 18 [*the date has been crossed out and replaced with 25*], 1851
Rose Spring, California
Dear Father,

"Hope deferred maketh the heart sick,"[29] indeed, and fortune seems at the present time to heap all her frowns on our poor devoted heads. However, we are use to it, so that it does not hurt us much.

After months of hard work we are just (almost) where we started from in our quartz hunting—no vein yet, with strong hopes, and assurances almost positive. And this is all we have to tell you.

Gold veins, generally, are very crooked, and our ill success has been owing mostly to this fact—our ground is very unfavorable for tracing them, and had half the amount of work been expended on the veins that we have spent in tracing and hunting them we would have a very good idea of their value. However, it is as it is, and we must be satisfied to take things as we find them, besides it is probably to our advantage, after all; with patience, and labor <u>we can</u> find them—very few can say the same. And what is more, father, <u>we will</u> find and they are well worth the finding.

Day before yesterday we sunk a hole on one, and a foot or so from the top of the rock, the quartz presented a very promising appearance—would probably pay for working. There is, we suspect near this hole a rich outcropping and we spent yesterday and the day before in attempting to strike the vein some thirty feet to the west—but it is so confounded crooked that though we have sunk a trench through the drift twelve or fifteen feet long across its course, it with our usual luck, has given us the "dodge." We are after it, though, and I think a day's work more will "head it."

Be not discouraged with our bad luck, father, for as discouraging as it may appear, our chances of success in our own immediate section on the veins we are and have been examining are as a hundred to one in our favor. Failing here, we have aboard, opportunities which amount almost to certainty—and by next spring at furthest, we will be in possession of more than one good vein.

I am sorry that Reuben and Warren cannot be furnished the funds necessary to come out here this fall, but it cannot be done <u>now</u>. Things now are so evenly balanced that tomorrow may place us in possession of means—or it may be a month yet. But the time must come and <u>soon</u>.

I write in haste, as you will see by the letter itself. We are kept so busy that the little leisure we allow ourselves, is not more than sufficient for rest. Yet a long and strong pull, and with God's aid we will be in smooth water—all of us!

To John and Prime I intended to write two weeks ago, but neither of us felt able. When we once <u>find</u> our quartz veins, they will have plenty to do—until then I must say to them as I do myself—Patience! patience!

Love to all.

Truly your son,
E. A. Grosh.

N.B. In taking up quartz veins Aunts Elizabeth, May, and Maria, as well as grandmother, Emma and the girls we will take the liberty of using their names to hold claims, presuming they will not object to turning gold mines as well as the male members of the family.

Allen

Our almanac got out of fix—Date this 25th.

We are all well here and love to all.

JG Jones

ABOVE: An ambrotype, a type of
photograph popular in the 1850s,
captured the image of Hosea Ballou
(*left*) and Ethan Allen Grosh (*right*),
presumably before they left the East
Coast for the California Gold Rush in
1849. *Courtesy Nevada Historical Society*

RIGHT: Reverend Aaron Grosh
(1803–84), the father of Allen and
Hosea, eventually became the librarian
for the U.S. Department of Agriculture
and a cofounder of the National Grange
of the Order of Patrons of Husbandry,
an advocacy organization for farmers
and ranchers. *Courtesy of Amanda
Brozana, the National Grange of the
Order of Patrons of Husbandry*

SAN

...he weak—my way and we ones, We arrived in Co...
...with three dollars in our pockets, and obtaining credit...
...procured mining tools, set to work, and made $3 or $4 f...
days. You can not imagine our vexation and chagrin...
...the first time we visited the bar on which w...
first worked,— all Cheesman's obstacles vanished— our...
...chine, with a few trifling alterations, would have...
to a charm,— It could not have failed! I cried fo...
vexation. Ho and I made a resolve, come what may...
to dig from the earth the funds to lay the foundation of...
fortune, and neither receive, nor solicit aid from a...
person.
    Not being able to do anything on the bar we mov...
...higher up the river, struck a rich lead, and for a whi...
...done very well. We worked...
...ated by a mining Co. to join with them, did so...

For correspondence dated January 5, 1851, Allen Grosh used stationery with a lithograph of San Francisco. Many '49ers sent this sort of letter sheet home to give family and friends a chance to imagine the marvels to be seen in the booming West. The artist, J. H. Pierce, created a variety of letter-sheet images of San Francisco. *Courtesy Nevada Historical Society*

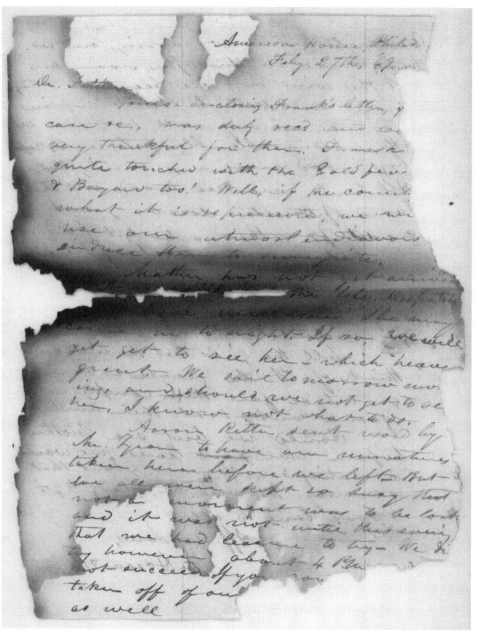

The Grosh brothers sent their first letter on February 27, 1849, from Philadelphia. The task of transcribing the deteriorated documents presented a challenge. In the case of this letter, obstacles included a torn right side, burning, and water damage. *Courtesy Nevada Historical Society*

On October 5, 1853, the brothers sent a letter from Lake Valley on the south side of Lake Tahoe. They were seeking better claims, ready to explore the east side of the Sierra.
Courtesy Nevada Historical Society

On September 7, 1857, Allen wrote to his father about the death of Hosea: "I take up my pen with a heavy heart, for I have sad news to send you. God has seen fit in his perfect wisdom and goodness to call Hosea, the patient, the good, the gentle to join his mother in another and a better world than this. In the first burst of my sorrow, I complained bitterly of the dispensation which deprived me of what I held most dear of all the world, and I thought it most hard that he should be called away, just as we had fair hopes of realizing what we had labored for so hard for so many years. But when I reflected how well an upright life had prepared him for the next, and what a debt of gratitude I owed to God in blessing me for so many years with so dear a companion, I became calm, and bowed my head in resignation. Oh Father, <u>Thy</u> will, and not mine, be done. Our happy faith in the perfection of God's wisdom and goodness will be your consolation as this cloud passes over your head, for well I know your heart is full of the great hope which caused Paul to shout in triumph, 'O death, where is thy sting! Oh grave, where is thy victory!'" *Courtesy Nevada Historical Society*

The tombstone of Hosea Ballou Grosh, in the Silver City cemetery. The community was founded in 1859, and the cemetery followed. Hosea's remains were moved to this location afterward. The tombstone subsequently fell to the ground, and locals encased it in concrete. *Photo by Ronald M. James*

The tombstone of E. Allen Grosh—he went by the name Allen—still stands deep among the trees of the Sierra Nevada near the abandoned site of Last Chance, California. Aaron Grosh, the father of the brothers, purchased matching stones for his two sons in the Far West. *Photo by Thomas Perez*

1852 AND 1853

Letter from Hosea Grosh:

March 17, 1852
Mud Spring, California
Dear Father,

Last Sunday I sat down and wrote a letter, but forgot it this morning till almost at town so I write a short one now. I will soon write again.

We [have] been as far south as Dry Creek 25 miles from our old camp but our old partner, Steele, leaving us there and buying into quartz vein, we returned, after staying awhile on the Cosumnes River. Steele turned horse thief about the time he left us, and only escaped [punishment] through Allen's intercession, who thought it a first offence, under temptation, and that Steele was sincerely repentant. Steele left the country immediately. Our loss through his rascality has not been much except in the chagrin and worriment the disappointment in one we thought a proved friend.

The season has been so dry that we were unable previous to the present month to wash more than 10 days at most and consequently fell considerably behind; but the present month will make all right as we have had rain plenty so far, and will be able to wash for a month to come. We are clearing about $32 per day with every prospect of holding out.

We will commence the summer season by one of us trying the river. The other quartz prospecting our hopes are as sanguine as ever and think we know where to strike the vein we want. Lack of funds and winter season have delayed us till now but no one can prospect before the waters dry up and rains cease.

Yours containing John Jones and Reuben Wells proxies were received also one letter dated in January but the letters containing the other proxies have not come to hand and probably were lost in the last mail of about that time. We don't need them now and if we want will send for them again. Heard from Dr. Mason but have not had time to go to see him. The weather is now fair again but not settled I think.

We are all in good health and hope that our repeated failures will not put you out of heart. I have time for no more at present.

Your affectionate son,

Hosea

ᶜ Letter from Hosea Grosh:

May 2, 1852

Mud Spring

Dear Father,

I had intended to have written before this but our ink was all spilled and I came to town but once during the past month and that soon after my last. I am sorry that we did not write oftener during the winter owing to lowness of spirits, etc., as explained in my last and will in future write at least once a month and as much oftener as I come to town. We are gradually paying off our debts together with those left by Steele. We could pay them at once by selling off the stock which was jointly owned by us but this would [be] a loss to us as it would be very difficult to supply them again. Our own debts are all paid but about $100, which is no hurry whatever as it can lay even for years without interest and without incommoding anyone but of course as soon as we can pay it without much inconvenience. Steele's creditors are with the exception of $300 in the same way which we will pay about 2 weeks the balance about $200 will wait without incommoding anyone until next fall and perhaps spring should it be of any advantage to us.

It seems hard that something should occur continually to set us back. In this case the fault lay in our being too anxious to get out and hiring hands counting for certain on water which did not come as the winter was very dry, but thank God the spring has made up for it. I think that before the month is out we will be clear of all encumbrances.

We have had threatening weather for the two or three weeks past which has caused lameness in my wrist, but except for 2 or 3 days, not sufficient to stop my working.

This morning coming over, my mule started and having an American saddle on threw me and jerked away spraining the middle finger of my left hand which pained me severely for a while but is quite easy now and I hope will not keep me from work.

I have got ink again and will write during the week and bring or send the letter early next week.

I have just received your letter mailed March 20 and am grieved at the uneasiness our not writing has given but can only promise to do better in future.

Your affectionate son,
Hosea

I would write more but have not time.

Letter from Hosea Grosh:

May 24, 1852
Mud Spring
Dear Father,

Though somewhat in a hurry I thought I would write a few lines. Shortly after I wrote last a man claimed one of our mules. The trial was held the Monday of the next week and laid over for further testimony since which I have not heard from him. The evidence at the trial together with an affidavit filed since prove conclusively possession of the mule at least three months prior to the time which he states to have gotten his mule and continued possession from that time to this which makes the mule so clearly ours that there can be no doubt of the result but it is very annoying and took considerable time. My finger is not yet fully well though so that I can work pretty well. I have been most of the past week getting in hay for the dry season.

Tomorrow Allen with two hands goes to the river to work and if it pays this week will go on. If not dismiss the hands and get along without them as well as we can.

As I wish to get home before noon I must now close. My health has not been good, though I have not been sick, since my last. Allen is in good health.

Your affectionate son,
Hosea

Letter from Hosea Grosh:

June 28, 1852
Mud Spring, California
Dear Father,

I have no time at present to say more than that we have just commenced work on our river claim and that our prospects are flattering. Neither Allen's health nor mine is very good though we are not sick. Last winter we worked too much in the water which is now being paid for. I will write, I hope at length, by next mail and close by saying we received your letter dated March 18th.

Your affectionate son,
Hosea

I forgot to say that 2 days ago I met 3 men from Utica, one named Glass[1] formerly clerk in Stephen Comstock's store, left Utica last February; wished to [be] remembered to you.

Hosea

~ *Letter from Hosea Grosh:*

August 7, 1852
Work's Ravine Mt Sugar Loaf
Dear Father,

I have so little inclination to write that I certainly would not do it but I know that otherwise you may think things worse than they are. We worked on our river claim . . . until we found that notwithstanding all our prospecting it was, likely to fail, we immediately dismissed our hands having sunk near $300 in the concern. We then went to prospecting . . . Cured the diarrhea immediately but the fever hung on longer and the . . . has not left me yet though I am now strong enough to be able almost to work and feel pretty well.

The claim is going on by the aid of three partners who we have in it and as soon as it pays more than $5 per day to the hand, we will realize something from it. We will do it we think as soon as they can get it fairly opened but with only 3 the work gets along very slowly and neither Allen nor I are able to work at it. Allen is not sick but he is not strong . . . We will . . . leave the claim altogether in their management and if it yields anything well and good; we will go on laying out our winter's work. We can pay current expenses and something more without working hard until we regain our strength when we will do what we can do. I think that when we have finished our winter's work we will realize over $5 for every day's work put on it for now and in the winter. This is cold comfort, but I hope it will not dishearten you, . . .

I remain your affectionate son,
Hosea

~ *Letter from Hosea Grosh:*

September 9, 1852
Work's Ravine near Mount Sugar Loaf, California
Dear Father,

I had thought to write by the mail of the 1st, but it was past before I was aware so we missed that opportunity so we write now. We have been busy throwing up dirt for a couple of weeks with pick and shovel after making several attempts make a plough work.

The great difficulty was the stones; after the first foot was thrown off they were packed so close that we could not get the plough to enter, we finally had a coulter with a point made which works pretty well and we think that by altering the point a little it will be just what we want.[2] Last week we washed some 80 pans of dirt (the best of course) which yielded $26 besides a lump which contained about an ounce of gold. This week we have washed so far about 30 pans which paid $9. We do this to pay current expenses, etc., therefore pick our best dirt for the purpose as we intend to wash with the tom and therefore it would be losing time to wash more than that. I suppose our dirt will average 4 or 5 cents to the bucket (about ⅓ of a cubic foot) besides lumps which may amount to as much as the fine gold or perhaps not more than half as much. We have one partner and think we can wash 1500 buckets every day we have water. We have one good dam in for holding water which will hold 2 or 3 days water and can by a week's work put in one that will hold more than 10 times as much so I think we will lose comparatively little time when the water does come.

Neither Allen or myself are fully strong yet but we think we have found out what ails us. A friend who is well read in medicine and has some experience, suggested that it was disease of the spine and said an external application along the spine of ointment composed of Tartar Emetic[3] and lard would cure it. By applying just often enough to keep up a slight irritation it is (he said) not in the slightest degree dangerous. We have tried it and find it to benefit us very much. Allen says that my spine is crooked; will you write to Frank and see what he says of our treatment of it?

I will give you, what I believe I have not done before, the geography of our location. Coloma our county seat is about 20 miles north of us coming from thence you pass through Cold Spring on Weaver Creek 6 miles further Mud Spring on the Hangtown road from Sacramento City.[4] 2½ [miles] further Log Town keep due south for about 6 miles on a ridge running along the Cosumnes River, you come to Work's Ravine in which we are the only settlers. Mt. Sugar Loaf is less than ¾ of a mile from us a little north of west. Quartzville on the Consumes, alias Nashville, lies due east 1½ miles. The road from Log Town is known by name of its terminus the Diving Bell Bar about 3 miles south of us on the Cosumnes River which there has just broken through the ridge before mentioned and continues its general course of southerly and westerly through every point of compass except perhaps east. Our weather for some days was very warm succeeded by 2 or 3 of moderate and now again warm.

11th, I intended to have written more but have been busy hunting one of our cows which strayed off and this morning Allen goes to town.

Your affectionate son,

Hosea

ఌ *Letter from Allen Grosh:*

October 11, 1852
Work's Ravine, California
Dear Father,

By this mail we can only send you a few lines—by the next we will try and write you in full.

Our healths are steadily improving—Tartaring our backs I think will pretty effectually right us. Hosea walks quite straight more so than I have known him for ten years.

Our diggings continue to hold out in their flattering prospects. We are sure of a good winter's work. We are throwing up [a lot of dirt]. The flat will yield us work for 2 years.

> Love to all,
> In haste your sons,
> E. A. and H. B. Grosh

ఌ *Letter from Allen Grosh:*

October 24, 1852
Mud Spring, California
Dear Father,

We have been very—very—busy this past two weeks, which must be our excuse for not writing you in full as promised last mail.

Our healths are still improving, and the flat still promises a good winter's work.

The mail is just up. I hope it may contain something for us.

The Whigs are very active and speak confidently of carrying the state for Scott. A democratic office holder has just complained to me of there being so many democrats in this section "tinctured with abolitionism." As he had just swallowed a bottle of Port, according to his own admission, I thought it best to sympathise with him and said it was very much to be deplored.

> Love to all,
> Your son, in haste,
> E. A. Grosh

ఌ *Letter from Allen Grosh:*

December 31, 1852
Glen Hannah,⁵ El Dorado County, California
Dear Father,

As you will see by the heading of this, we have given the name of our dear mother to our place and opened a ranch—or rather, are opening one—for we have not yet perfected our arrangements.

Yours of November 17th (postmark) was received last Monday, and was very welcome. The "Manual" we had already received, and read hastily, when we handed it over to Mr. Disheroon, an Odd Fellow and Mason, who was very anxious to examine it.[6] We were very much pleased with it, as was Mr. Disheroon, who, when we last saw him, had given it a hasty look over. We will probably get it back next Sunday, and also his opinion and that of several other leading Odd Fellows, to whom he has showed it. Mr. Disheroon, is one of our warmest friends, and entertains a very high regard for you. He wishes us to remember him to you, and has, with us, some hope that you may yet make California your home. He expects his family out here next February. We are rather surprised at the quarter from which you experienced opposition to the Manual—and can attribute it to nothing but a spirit of Hunkerism.[7] Is it not strange that most all successful scholars, dim the glory of their first triumphs, by turning their arms against those who in their turn would keep abreast with the "Spirit of the Ages." It may be human perfection cannot be reached by the present age—but until it is reached, nothing can be permanent. All we can do then is to hold on to that which we think is good until we find something better. "Prove all things—hold fast that which is good,"[8] seems to be disjointed— two parties—each preaching from one-half the text. In this strife, you, father, seem always the same—avoiding both extremes, swayed neither backward nor forward—steadily, surely, advancing, never laying down a principle without a recognized base—never stepping without a principle for a standing place. This is the breathing spirit of your book—and as such, we are proud of it. In every other way it is worthy of you.

You chide us for not making some arrangements for coming home, even if they are distant. For the coming year, it is impossible to think of it. We have given up mining—we cannot follow it without prejudice to our healths— and taken up a business more agreeable, healthy, and, I think profitable— ranching stock and farming. For the former we have an excellent location. We have thirty-odd head, now, before we are ready to take them. In a week or two it will be up to 100 at $2 per head. This branch of the business takes up but a small portion of our time. We will put in considerable grain, and next spring, as large a garden as we can manage with the help of a hand or two.

We will put in considerable corn (sowed) for fodder, and our friends think by next spring we will run up the stock to 300 or 400 head. After next August we will increase the price per month to $3 or $4, and feed—we can do it so much cheaper than any person around here that we will probably have as much as we can tend to.

The winter has set in with all the severity of '49. The mines are destitute of provisions. Hosea goes to Sacramento, tomorrow for flour, etc., for ourselves and neighbors. He packs, with 4 mules, at 15 cents per pound—45 miles. If it proves profitable he will follow it up.

Our claim we have made arrangements for working, and I think we will get something from it. It pays very uneven if it fails we are not responsible for any loss.

We lost our two cows and a calf by theft about a month ago—they would now be worth probably $300 or 400 if we had them.

I must close abruptly Hosea is going.

Affectionately your son,
E. A. Grosh

✑ *Letter from Hosea Grosh:*

March 6, 1853
Glen Hannah
Dear Father,

Having an opportunity to send a letter so as to go by first mail I write a short note to let you know we were yet living. We will write more fully next time. We tried ranching stock but it did not pay of itself and kept us so busy that we had time for nothing else.

Though hard work, it improved our health. We needed a little rest after it. . . .

We are going to gardening and expect to do well. We will be able to judge better next time we write,

In haste your affectionate son,
Hosea

✑ *Letter from Allen Grosh:*

March 26, 1853
Glen Hannah, California
Dear Father,

We have received no word from you for some time, but fully conscious that we deserve it, will find no fault. Our hasty notes and ill-fulfilled promises could find no excuse, were it not that we are kept so busy climbing the hills of adversity that we are continually out of breath. This past month we have been resting—casting everything aside, we have "taken things easy"—hunting, working when we pleased, etc. We are now ready for another start . . . to be out of breath when we enter the race, and the next month, we will jog along at the same easy gait we have been going this, when we will pass to the east side of the Sierra, and be that much nearer home at least. In other words, we have what we think good and reliable information of diggings near the summit— east side—above the Carson River valley—should they be what we think they are, there is but little doubt of our doing well, either at mining or something else—mining <u>only if it is sure</u>. Failing there, we will go on the river, engage in what we think most profitable, make what we can, and send it home, and if it is enough to satisfy us go along with it. Since we left you in February '49, we have had but one object and one aim—independence and happiness for the family—and in such a manner that we might enjoy them together. Even after your refusal to settle in this country we struggled on in hope that in the event of our succeeding you might be induced to alter your determination. We failed—so let it go—it is probably for the best. Hereafter we will take a different method. When we have anything to send be it much or little we will send. We heartily approve of your plan of securing a homestead, and will do what we can to bring it about. You can invest it there, or elsewhere, as most gives <u>you</u> pleasure—if we ever get any to send. At any rate, if we do send, it is entirely at your disposal. We will try also and aid you in giving Warren[9] his medical and Malvina her musical education, and we think we can. Had we adopted this plan at first we would by this time have contributed considerable aid, and though it is a rule with us to do the best we can and regret noth- ing, still a hang will often arise that we have been of so little service to you when we might have been of so much. We have neglected you father, and said but little, yet you, and all, are constantly in our minds. Uncle Pinckney has our warmest sympathy. Hosea worries himself much about it, and would do almost anything to aid him. But there is one thing—we will not go home until we have something to go home with. We have trained our minds to it, and steeled our hearts to it. We can do more good out here. The life is no lon- ger one of hardships to us and our healths for money is plenty and will be for some years—our bad luck—for <u>luck</u> it is—cannot last forever—tiempoco[10] as the Spaniards say. This much we promise you; prudence in all concerning life and health. Neither will we hazard by exposure or imprudence.

We had intended to go into gardening here, and had matured several plans for speculation in an agricultural line, but we anticipate almost universal fail- ure among the miners the coming summer, and they will carry everything with them. Accordingly, we have concluded that it was best for us to stand clear—our luck has never been any of the best. This winter has been a very hard one on the miners—very few successful. We have concluded it best to be off before the final crash comes. If we find a good opening in Carson Valley for a garden we will give that business the preference.

A word about the homestead. If we succeed in making considerable the coming year, if you think well of it, we will come home via the Plains bringing with us a <u>manada</u> of Blood California and Sidney Mares[11] (the best and finest in the world, the last well suited for the Pennsylvania market.) a stud of well picked Jennys for raising Jennetts (the prettiest and best riding animal in the world) and a couple of jackasses for mule raising and also some of the last for trading with the Snake Indians[12] for a stud of their mares, swift, beautiful, kind and gentle. The business of stock raising would suit us, and we think all much better than anything else. A few thousand invested here would make a magnificent start with you. Please let us know what you think of the idea and we will write our views more fully at a future time. Please consult Pinckney and the rest. Love to all.

Affectionately your sons,
Allen and Hosea

Mud Springs will remain our post office.

❧ *Based on handwriting it appears this letter is from Allen Grosh:*

May 22, 1853
Glen Hannah
Dear Father,

We received yours of March 2, today, and sat down at once to commence an answer, though we intend to take all week to finish it. We have little to write concerning ourselves that has not been said before.

We will leave for Carson Valley as soon as we are certain that the mountains are passable for animals; the rains continue so late, and are still threatening, that there is little hope of this being much, if any, before the first of July. We are selling off everything we don't want to take with us, as fast as we can; and sacrifice but little on anything, as our time is sufficient for us to do so.

We will get early information in regard to this by the Salt Lake mail. Large numbers are going over to purchase cattle, so that there will be no scarcity of company; indeed, except at first, there is no danger except to small parties, but the snows continuing so late . . . We have received information of diggings which if correct we will do well in: our information is as reliable as can well be. The openings for other business are better than here, the country pleasanter, the soil better, more chance for water for irrigating, and the climate fully as pleasant and healthy as this, and our continued ill success has sickened us of this country. You need not fear that we are going to ramble about much, for we will settle on some place that suits us, and stay there, until, we leave for home, sweet home. We will, if we do well at all, probably come to see you next spring, but hardly hope to be able to stay there, as we will not live in the old states poor; but should fortune favor us with sufficient, we will settle down somewhere with all the rest if we can possibly bring them all together. May God spare us all till then is the constant prayer of our days. You are, I see, 50 years old today, (if I mistake not, and it is near time that you rest, from the toil and trouble which have been heaped upon so many years of your life, should begin.

We botanize and gather flower seeds during our leisure. In this letter we will send some of the fruits of our industry in that line. Tell the girls to be very careful in opening the papers marked "minute" as the seeds are exceedingly small. We will mark each with its generic name, when we can find it out, which in many cases we cannot; the others with the name of some flower it looks something like. They may not look much like they do here however, as soil and situation make a great deal difference in many of them. The great feature in this country is not so much in the variety (though that is great) as in the immense numbers of each kind forming in patches of different colors, so as to give a peculiar appearance, well described as "carpeted." The larger and more brilliant ones are generally scattered through the other less important ones, which strike most by their number, so as to add wonderfully to the magnificent scene. Such ones as are likely to become troublesome as weeds (which are few) we will particularly notice so that care may be taken . . . . in our next, you will get others. . . .

24th. Our old partner Williams has gone to Australia and I am not sure but we might have gone with him but for the folks at home. Our news from there is of increased yield in gold and trouble with a prospect of increase in the latter at least. The governmental regulations are very strict and are made more onerous by the petty annoyance and overbearing conduct of many officials. It would not be surprising to hear soon that a revolution had commenced there. Indignation meetings in which Australians and Americans take part remonstrating energetically to the government. What notice will be taken of it remains to be seen.

Placers as soon as discovered are laid out in claims 8½ feet (more like a ½ rod) square,[13] and any man to work must take a license for working a particular claim: if caught working or prospecting on any claim without a license or on any claim except the one for which he has a license is fined $20 and in default of payment committed to the chain gang for as many days.

The social condition is bad decidedly. A miner cannot leave his tools [or] anything else without it is stolen. After they get down in their holes (the diggings are from 12 to 50 ft deep and the bottom only paying) [they] dare not leave them even at night or leave any dirt outside them for the same reason; bushrangers[14] abound in every section. Everything has all the worst features of California in '49, and worse, without its redeeming points. We will receive news more reliable from there, soon after Williams's arrival than is to be gathered from newspapers or scattered reports; they will [come] from one on whose judgment we can rely, who will be an actor in the scene. We will write to send us flower seeds if he has time.

In washing there has been a decided improvement made, in this country, by using quicksilver in the sluice. The sluice is a succession of straight troughs

12 feet long (we call them boxes here) 12 inches wide at one, and 14 inches wide at the other end; sides about 10 inches high. About a foot from the lower end a cleat 2 inches high is placed, which in the lowest that dams back water about 3 feet the others a little less. . . . a large head of water is used. The quicksilver is applied by taking about ½ a pound (for a 100 foot sluice) in a handkerchief and squeezing it the whole length. The operation is repeated about 6 times a day, so as to keep some quicksilver in sight the whole time. It is said in common dirt, to increase the pay from 50 to 75 percent, in some, much more. It will probably come into general use during the coming summer; if so, more money will be taken out than any season since the mines were opened. We might possibly stay, and try it on the river, but our healths have much improved during the past winter, and are still improving, we therefore do not like to risk it again where we will be in so much danger of sickness. Health first, other goods after. We will take quicksilver with us, and try it the other side of the mountains, if we mine there.

Our seasons here seem to have changed entirely. Our rains in March were light but they have continued to the present time and there is one now threatening. Is our climate to change? The ditch for conveying water and the quantity of soil turned over are generally thought to be the reason. If so we are likely to have rains like with you.

June 12.

I have little or nothing more to add though it is three weeks since I commenced. Enclosed you will find what seeds we have gathered so far, these are of our earlier flowers and will therefore be more likely to succeed in ordinary soils and situations; with the poppy you had perhaps plant some in your warmest exposures. The clover grows on our hills in preference to moist places and may be valuable for raising in poor soils and dry situations. Perhaps the flowers of every kind, especially those we send hereafter, may succeed better in Pennsylvania than in New York.

We are both now in fair health. The weather seems now to have settled down at last.

We will write again just before we leave and also as soon as we arrive at our destination.

. . . Allen and Hosea

✐ Based on handwriting it appears this letter is from Allen Grosh:

July 13, 1853

Dear Father,

We will start for Carson Valley tomorrow morning. It is uncertain whether we stay there or not. Some of our friends are anxious to have us stay here, and if we cannot do <u>very well</u> on the other side of the mountains we will return in the course of a month. If you do not hear from us in that time you may count on our having "struck something" handsome. If we return here we will . . . both home next spring on a visit.

Our healths are much improved and improving.

"Ho! for the Mountains!"

Love to all.

In haste.

Affectionately your sons,

E. A. and H. B. Grosh

P.S. We send a few more flower seeds for the girls.

∾ *Based on handwriting it appears this letter is from Allen Grosh:*

July 31, 1853

Mormon Station,[15] Carson Valley

Dear Father,

We arrived here, after a very pleasant trip, yesterday night, all well.

We will tomorrow, start again, and I think we will find something worthwhile.

Diggings have been discovered on the East Fork of Carson River, and if we cannot do better we will go there. I don't think we will go back to California. We will not follow mining longer than until. . . .

Our kindest wishes and warmest love is all we can send you now, but will write at length within two weeks, giving a detailed account of the journey and country.

Affectionately and truly your sons,

E. A. and H. B. Grosh

∾ *Based on handwriting it appears this letter is from Hosea Grosh:*

August 17, 1853

Lake Valley[16]

Dear Father,

We have just this moment returned from a pleasant though rather rough trip among the mountains—well and hearty. Do not be alarmed at reports of Indians being bad, etc. Watchfulness and care are sufficient to guarantee safety—we neglect neither. We start out again immediately with tolerable fair prospects of finding something handsome. Excuse us not writing [at] more length. Love to all.

In haste truly your sons,

Allen and Hosea

Based on handwriting it appears this letter is from Allen Grosh:

October 5, 1853
Lake Valley, Utah Territory
Dear Father,

We have for the last six weeks been prospecting in this valley, but have not struck anything worthy [of] note. Our hopes were to do a little something before we wrote, but they are gone for the present, and fearing that you would feel uneasy about us, we write now.

We will leave here in a day or two for the mountains east of Washaw Valley,[17] which is north of Carson Valley. If we find nothing there we will cross to Gold Canon, which empties into Carson River, and in all probability winter. . . . The miners here are about two to three years behind the age, so one acquainted with the machinery now used, for gold saving, in California has at least 5 chances to their one, of making money. In fact he can give them as wages to work for him, as much as, they can make and then clear at least as much more on them. Where we are now is the place to spend the summer we ever saw. It is cool always in comparison with California cold, but so equal in temperature as to be exceedingly pleasant. The lake in this valley (for there are three, some say connected) is 40 or 50 miles long and about 25 wide as near as we judge by seeing in different points. It is certainly a beautiful sheet of water surrounded by mountains said to be the head of Truckey River but it looks otherwise.[18]

We will not write in all probability for about a month when we will go to California to lay in our provisions.

With love to all we bid you farewell.

Your affectionate sons,
Allen and Hosea

P.S. We have found several opals,[19] one we think is valuable. . . .

Letter from Hosea Grosh:

December 3, 1853
Gold Canon, Carson Valley
Dear Father,

We have not for some time been able to send a letter across the mountains, and did not expect to, for near two weeks, but an unexpected chance just offering, we concluded to send a short letter for fear of accidents.

We have done little since we have been here but now we have our sluice going and we will do something soon. As it happened we have had some considerable quantity of poor dirt to wash. . . . Our health has generally been good but yesterday I had an attack of cholera morbus[20] or, something of the sort, and consequently have not been working today but will do so tomorrow.

We will write our <u>tour</u> through the mountains next Sunday. Would have done it before now but getting up a house, (which is considerable of a job where there is no timber) and getting fixed, etc., trying to make expenses has kept us pretty busy besides, not having the materials for writing I had to go to a neighbors to write this.

There is certainly the best chance to [do] well here of anywhere I have been. Provisions are not as dear as we thought they might be; though I suppose you will call them high. Flour 35 cents per pound, potatoes 10 to 15 cents, turnips 5 to 8 cents, beef 25 cents. Our living cost us from 75 cents to $1 apiece.

Allen is waiting for me and it is dark so I must close for the present so farewell.

Your affectionate son,

Hosea

P.S. I forgot to say that we will probably have a monthly express across the mountains so we will have an opportunity to send and receive letters. We will at least have one a couple of times during the winter.

1854 AND 1855

❧ Letter from Allen Grosh:

November 8, 1854
Little Sugar Loaf,[1] El Dorado County, California
Dear Father,

We are once more in California, after passing through trials and hardships, which five years ago would have sent us to our graves, and I am happy to say our <u>healths are entirely reestablished.</u>

We find things here but little altered, and we have every prospect of a profitable winter's work before us; in which case <u>Hosea</u> will be home in the <u>spring</u>.

We are engaged now in a thorough, complete and systematic "prospect" of our section, and will continue it until the rains, and we think, that in another winter we will have our "piles."

We found on our arrival here a letter from you enclosing one from 'Vine, and another from 'Vine. They were welcome treats! We will send this to town by the first opportunity, and follow it up by answers to your and 'Vine's letters, and by sketches of our life in Utah, as fast as our leisure will allow us to fill them out. If we have disappointed you in our promised visit, we make what reparation we can by a diligent correspondence. <u>Warmest love to all!</u>

Truly and affectionately your son,
E. A. Grosh

❧ Letter from Hosea Grosh:

February 7, 1855
Little Sugar Loaf, El Dorado County, California
Dear Father,

We have been trying for some time to get a letter home. Allen has tried and tried, and finally calls on me to write a short letter at any rate, and he will then write a long one. As half a loaf is better than no bread, I try my hand. Things have been very dull here as all over this state. We have been getting along and that is all. The great bulk of the gold since '49 and '50 has come from the dry diggings. Everything has waited for and been dependent on them. For several years they have been the wheel greasers of commerce; this year the rains so far have failed almost wholly. As far as mining is concerned, they did not even start the ravines in this section, and did very little good anywhere; so the aforesaid wheels, for lack of greasing have nearly come to a standstill. For a week or so past we have been doing a little better, averaging nearly $2 apiece per day. We are only able to work half the time for lack of water, so that when it comes we will be able to make at least $4 dollars per day. At present we carry our dirt a short distance and will probably continue to do about as we have for the last week, until water does come, when our wages will at least double, as the "packing" will be saved.

The miners are not alone in their trouble. The farmers have been delayed in their plowing and on account of the hardness of the ground have not done it half at that; the general yield therefore will likely be less even though we have a fair season, than usual. There is perhaps more ground put in, which may bring up the aggregate as high as last year. We fill up our leisure time fixing around the house, making moccasins out of boot legs,[2] and everything we want, as far as we can. This saves in expense. . . .

Hunting fills up the balance. We are commencing a garden so as to raise our own vegetables which will add much to our comfort and be a great saving as it needs none but spare time to attend to it.

The quartz vein we looked so much for has been found and we have 3 claims on it. At the place it has been opened it is rich; the commencement of our claims is 125 yards south from this place and as no prospecting has been done on them there is no telling how they will pay but we hope well. It is probably the vein which has thrown out the gold on the ravine we formerly worked on. Owing to the dryness of the season, we will not attempt to work [the] best placer claim, as it lies on the side hill and the only chance is working during and immediately after heavy rains. We will therefore depend on the claim we are now on as the surest. Missourians and Kansas[3] I see by the papers are having trouble; one would think that the result of "squatter sovereignty" would as far as developed, satisfy anyone. Congress alone should settle the slave questions as regard the territories. As for the Missourians, it is just what we should expect from them, from what they did here in '50. They went in "flocks" and tried by superior force to frighten and bully the timid. Like the ghost of the Highland Chief in Scott's St. Valentine's Day[4] a firm resistance always drove them off, for real fighting was too dangerous to suit them. They are now nearer home and may do more than they did here, but I think that like here the name of Missourian will remain a word of contempt.

The [Crimean] war in Europe excites considerable interest though I believe the general opinion is that so long as it is a war in which the masses have no interest any very active sympathy is more than we can consistently give. Meanwhile we <u>have</u> a question of the People vs. the Aristocracy[5] to settle at home; and it had better be done soon or the Kansas and Missouri troubles may even breed civil war, especially should the wise heads of the South undertake to encourage, support and aid Missouri.

In this state politics are bad enough though on the whole rather improving in the undercurrent. Our principal candidates for senator are I believe all duelists and by our State Constitution disqualified for any office of honor, trust, or profit.[6] This is a pretty state of things truly, but the very greatness of the political corruption here, may bring about its remedy. The last move is that the members of the legislature voted to themselves one thousand dollars each cash to be paid them out of the State Treasury as part of their pay. Now as this would take all the cash out of the Treasury, and therefore oblige Governor Bigler to take his pay in script, he vetoed it; but the Legislature not to be outdone passed it by over a ⅔ vote.[7]

It has rained a little all day today and though it will not in all probability help the miners much it will the farmers. You can almost see the grass grow. Everything is beginning to look green for the first time this season. We have some flowers in bloom; I have seen buttercups, lupines, wild cucumbers, and 2 others, for which I know no names; beside the manzanita, which has been blossoming all winter, and is now dropping its blossoms; but the grass did not get start sufficient to carry it through the dry weather we have had, but has gone backwards rather than forwards. The present warm rain is doing much to remedy this; and in a few days all the hills will be green, and the grass highly approved by horses and cattle, who are, or ought to be, competent judges.

The present subject exciting public attention below, to judge by the papers, is a stage road across the plains, and different routes across the Sierra. From the different accounts, making due allowance for each being by far the best route, and other exaggeration, one thing is pretty clear; that it is a matter of no great difficulty to find fair, or even good routes or what might be made such by the expenditure of $50,000 or less. Our experience gives the Johnson Cut Off,[8] by far the preference over the Old Emigrant Road, and it is probably as good if not better than any route at present opened, i.e., it can be made as good a road as cheaply. The principal causes of its not being better known are the wholesale lying of its openers, causing the public to doubt their whole story, and there being more interest concentrated on the old road.

But we feel confident that the best route is not yet known at all. We have made inquiries in regard to the country lying between Sacramento City and

Humboldt River; and from what we can gather, and what we know of the country ourselves, we are confident that a road can be constructed which will be almost a straight line between these points, which will be of an astonishingly regular grade and cross the mountains at as low if not a lower point than any now known. If we have leisure the coming summer we will go over it and see what it is like.

Give love and respects to mother and the girls.

Your affectionate son,
Hosea B. Grosh

ঝ Letter from Allen Grosh:

April 6, 1855
Little Sugar Loaf, California
Dear Father,

We have sent you by Thomas L. Disheroon Esq. our opal. He sailed from San Francisco, last Saturday by the steamer for New York. Should he there ascertain it to be <u>very</u> valuable he will probably deliver it into your own hand, as he is very anxious to see you and make your acquaintance. We hope he will call on you at any rate, as we know he will be pleased with you and you with him. He is about your own age, a warm Odd Fellow and Mason, and has been to us a warm and true friend when we most needed friends. Should he come please give him a warm reception, and I know you will not regret it.

As to the stone, we have so far been unable to get any information whatever in relation to its value but our desire is should it fall under $1000 to keep it in the family. Should it go over that, we place it in your hands for disposal, just as if it were your own—If you think it best to dispose of it immediately, and at a sacrifice—do so—or hold on to it for better times—<u>just as if it were your own</u>. Only, give us an early information of its value as you can. For should it prove valuable, we will spend a month or two looking for another.

Now do not think from what we have said above that we have built large hopes on this stone. Far from it. If it is worth $5, well and good—if $50,000, so much the better. It can hardly rise so high or sink so low in value as to <u>astonish</u> us.

We have done very, <u>very</u> bad this winter. Bad luck is at our finger's end. Though we have not worked very hard, yet we have worked steady day after [day] for 50 cents to one dollar, and even less. This mining is a strange business. . . . The greatest difficulty we have had to encounter since coming over the mountains is "false prospects"—we can hardly take a pan full of dirt, without getting a good prospect. But the moment we set our tom or rocker the gold seems to vanish—it's not "thar." We once before had such a "streake" on us, for 5 or 6 months when our luck suddenly changed, and we could go to work almost anywhere and make money—$8—$10 and $20 per day—perhaps this run will take the same turn. I sometimes fret a little, but Hosea takes it as philosophical as though he was a party unconcerned—"only a passenger." Patience—patience—and all will come out right yet.

We received last week, one of your letters of last fall—by a mistake of the postmaster it was placed among the advertized letters—giving us the long merited scolding which we have so long and so richly deserved. Now father, in trying to write that "long" letter we have neglected the short ones—and to prevent the occurrence again—we will at once put off the long one until good luck smiles on us again. It is easy enough to dash off a page or so under almost any circumstances. But when we are compelled to practice the most pinching economy, as we now are and have been doing all winter, and can meet misfortune and bad luck piled up, with nothing but patience, and resignation—a firm fixed faith that all is for the best—why the less we think of home and the dear ones there the better. Our foot is on the burning plowshare and we <u>must</u> and <u>will</u> pass over. It <u>must</u> and <u>will</u> come out all right—I feel it—and then we will not only look back but <u>come</u> back. When that will be God only knows. We will do the best we can and leave the rest to Him who has so often guarded and feed us almost miraculously.

I write now on at least a quire of fragments, and scraps[9] of letters to you, the girls and Pinckney. On the first dawn of a better time we will see what we can find among them worth the calling. Until then we will send <u>short</u> notes. I do not think, dear father, that our long separation has in any way impaired our love and affection for any of the family. Our hearts now beat as warm at the thought of <u>home</u> as it did at the end of the first year. But have both contracted the habit of throwing off our mind from any subject when it becomes painful. Had we not done so, we would have sunk into our graves long ago— Our <u>healths</u> require this. The wear and tear of mind this winter has, we found lately, affected us. We at once threw care to the winds, resolved to do our duty, work patiently steadily and carefully, and <u>wait</u>. We <u>won't</u> worry, and things go smoother. We have both altered much in our manners. We are both as cool and steady and as unexcitable as two Russian veterans. Hosea thinks we are trained for some special object. I laugh at him; but he insists that Providence don't "put" folks "through" such a thorough course of training without giving them work to do in the end. All I hope is that it may not be gold digging. I think we have had enough of that—unless our services should be better appreciated.

Tell 'Tish she must save a <u>little</u> piece of that wedding cake anyhow. We thank her and 'Vine for their letters—and had not I made a firm resolve not to make any more promises I would say we would answer them <u>soon</u>. Our kindest love to Emma, John, and Cyrus and mother,

Affectionately your son,
E. A. Grosh.

P.S. Our habits are more regular than they were at home—California has rather improved than injured our morals. I mention this as you seem to be uneasy in the mislaid letter. "A man in California either grows better or worse" is an axiom here. I think I may say that we are either in the first or an exception to the rule.

Letter from Allen Grosh:

April 21, 1855
At the Old Place
Dear Father,

We received yours of March 5th, last Saturday. As you will see by the heading of this we are back again to our Old Place.[10] We moved up here last week, and are now living in a tent. We have had two rains of three days each since, which has prevented us putting up a house, and must excuse the shortness of this note, as during the rain we were obliged to seek shelter from a neighbor who already had four persons in his house and water has been so scarce this winter that we cannot afford to lose any of it while it lasts.

The water has given us an opportunity of prospecting our best claim (mostly on the side hill of a small dry ravine) and it promises well. We expect one more rain, which will enable us to finish our prospect of it when we will know if it will be as valuable as we hope. We want to throw up the dirt during the coming season, and sluice it through during the rains of next winter. We have a sluice-head of water only for a day or so at a time, during and immediately after a heavy rain.

A friend, Theodore Daft, has given us a claim on the Old Creek (known as <u>Grosh's Ravine</u>) on which we can make from $1 to $4 per day each, <u>certain</u>, with a fair chance of doing better. So you see our luck has changed, and better times are before us. But we have had a very hard time of it this winter.

Our health is not as good as we could wish, owing doubtless to the worry and trouble of the past four month, and the bad water at the place where we lived during the winter—Little Sugar Loaf, a mile from here. But as quick as the water gives out we will take a trip into the mountains, and two or three weeks will fetch us all right again. Hosea has had his rifle cut off so that the barrel is now only 31 inches long. He goes after it tomorrow, when he will mail this.

You mention in your letter the rumor that Warren contemplates marriage

with one of Mr. Libhart's daughters.[11] The high respect which we both entertain for Mr. and Mrs. Libhart would make us extremely happy in thus connecting our family with his more closely, were it not for the extreme youth
of both parties. What is the use of all this haste? We have seen so much of
these "youthful marriages" in this country that we are <u>very</u>, <u>very</u> sorry—nay,
we tremble—to hear of the step he is about to take. If they really love each
other, let them show their affection—not by a hasty, precipitate marriage but,
by patiently preparing themselves for the position they wish to assume. A year
or two of careful thought, experience and observation, may completely revolutionize their conceptions of the married state, at any rate they will at least
be the wiser, for the delay, and it can hardly do them harm. But I must close.

  With warm thanks for the many kind expressions of love, you send us from
all, we are

Yours in haste,
E. A. and H. B. Grosh

~ *Letter from Allen Grosh:*

May 18, 1855
Grosh's Ravine, Mt. Sugar Loaf, California
Dear Father,

  We received yours of March 30 a week ago last Sunday, by the hand of
our friend and neighbor, F. J. Hoover Esq., or as he is universally called <u>the
Governor</u>.[12] He is from Maryland and a relative of ours from that state, David
Clingan, came to this country with him in the year 1849. David is at present
a resident of Marin County which he represented in our legislature of '53 and
'54. We think much of him.[13]

  Your letter was read with much interest. We feel very sorry for John in his
trouble. By what you say we take Este to be his brother-in-law? The explanation regarding Warren's marriage makes us easier, for we both were afraid
that he was about taking a step which might give him much unnecessary trouble in future. We regret very much his misunderstanding with Aunt Elizabeth,
and hope it will be of but short duration.

  Your dyspeptic pains, dear father, we feel certain you can remove by the
use of the ointment the receipt for which Hosea sent to uncle Pinckney last
fall. That either of us are alive at the present moment we owe to it.[14] And if
we had used it vigorously from the first we would have been saved much suffering. Neuralgia, headache, jaw-ache and I know not what all accompanied
our dyspepsia. We were very low indeed when the Governor first urged its use
upon us. And though we have since passed through hardships and exposure
enough to kill 19 out of 20 well young men from the states we have gradually

been growing stronger and healthier. Hosea has had two backsets this winter. The first occurred early in the winter. He fell while we were carrying dirt in a handbarrow, striking the lower part of the back-bone on a pointed rock. The pain as you may suppose was awful. In a couple of weeks, his whole nervous system was affected. By keeping his back well broke out he was fully restored in about six weeks. Some three weeks ago he fell, again striking the same place. The blow was not near as severe as the first, but he commenced sinking rapidly. He is now getting better, and by the first of next month will be able to work again, as well as ever.

I cannot say that I am perfectly cured, for when it is cloudy I am inclined to dyspepsia, head and jaw-ache. But in twelve hours it will remove all symptoms of pain even at those times. The moment the back breaks out I have relief. The same can be said of Hosea, except during the time he has been suffering from the effects of his falls. I think that by knocking about in the mountains for a season or two our healths will be entirely restored. The pure air and water, and the hearty food and exercise will, we think, work off the remnants of disease and bring our whole systems into healthy action again. We are both much straighter than we were at home.

Sunday, May 20th, 1855.

The ointment is made by mixing as much Tartar Emetic as will lay on a dime with an even teaspoon full of lard. Anoint the back bone from the junction of the spine with the skull down its whole length, with a piece or lump of the ointment about as big as a pea, in the evening—repeat next morning. If small pustules do not break out by the next evening repeat until they do come out. If applied too liberally there is danger of producing ugly ulcers—but of this there is not much danger with ordinary care. Should it become sore, wash with castile soap, or what is better, a little sweet oil, or lard. The object to be attained is to keep up a gentle eruption so as not to be of much inconvenience—the extent of which will be a little itching. When the eruption appears stop anointing until the pustules commence drying up. Then anoint again. On some persons the affected parts of the spine break out more readily than those that are sound. In both Hosea and I the reverse is the case. Observation will be your guide in this. Use it one month, and you will never be without it. It may take years to effect a permanent cure, but it may be discontinued and recommenced at pleasure. After using it a month or two it will give you relief in 8 or 12 hours, always. Of course the closer you follow it up the sooner it will effect a radical cure. I could give you instances in which confirmed invalids have in a few months been made stout men.

The miners of this section, and indeed, so far as we can ascertain throughout the whole state, have had hard times this winter. As for ourselves we have been unlucky among the unlucky. Although we have practiced the most rigid economy and tried hard to make our expenses we have fallen behindhand nearly $50 since coming back from Carson Valley. Hosea does not work now owing to his back, but takes care of the house, and hunts, and in this way makes as much of our living as I do by milling. The past week I averaged 50 cents per day! But we take it coolly, however, keeping our minds as free from all trouble and vexation as possible for while we can get enough to buy flour, coffee (an article above all price to teetotalers in this country of hard knocks and blows) and powder, we are in no danger of starvation, and as long as we can keep that at arm's length we shall not worry ourselves.

From the last injury Hosea has suffered considerably but he is now improving and I think, without doubt by the first of June he will be hearty again. As for myself, my health is not very good. The continuance of cool wet weather so late in the spring (it is raining now) has together with the hard water, made me bilious and I can't "get right"—besides I was fool enough to allow our continued bad luck to worry me during the middle of the winter. Hosea was more philosophic—"We can't help it," he would say and that was sufficient. Hot weather, or a trip among the fishes, ducks, geese and deer of the mountains would fetch me right in a few weeks.

We are now waiting to hear what word you send as of our stone. Should it be very valuable we shall make what arrangements we can for our provisions during the summer and hunt for more—not that we hope to find another like it but, others of less value. Should it be worth less than one thousand dollars we will at once start into the mountains, and spend the summer there, living on fish and meat. We have a claim here on which we can probably do tolerably well next winter, but it is situated so that we can only have water during the rains. By going into the mountains, we can make an easy living (which we are really afraid we cannot do here during the summer) enjoy ourselves, and return next full, stout and hearty to work it.

"Uncle Tom" Disheroon thought that if our stone did not prove very valuable he would send it you, and not call on you personally. Whatever may be its value, we hope you have seen him. We have been acquainted since the spring of '51 and he can tell you more than we can send in a dozen letters, in a half-hour's conversation. His blunt honesty, and unflinching regard for truth, have, I am sorry to say been much injury to him as a trader. He hates anything like a "sham" and whenever he detects one he is very apt to speak out, just what he thinks. He has many warm friends here and is much esteemed throughout the country.

You complain of our letters being short and unsatisfactory. Heretofore we wrote that we had done nothing but hoped for "a better time coming." Now our hopes and prospects are summed up in fewer words—"Done nothing and

don't expect to"—and if you can give us directions how to make it more full
and explicit we shall be glad to receive instruction. We have <u>tried</u> to do better
than we have done, of course—and I may add that though there are many who
have not as near paid their current expenses as we have, yet there are but very
few but <u>might</u> have done so if they had <u>tried</u> as hard as we have. But indeed,
dear father, when everything goes wrong and nothing right—when you find
that everybody around you are making two, three, and four dollars, and can't
find a place in which you can make more than fifty cents a day, a man has his
hands full in keeping his temper in tune. So for goodness sake don't ask us
for long letters. I have written a sheet full and this, but have spoiled at least a
half-dozen besides.[15] And yet, I dare say you will find but little mark of deep
thought or careful composition in it! When I am blue, I grumble so much that
I am ashamed to send it. When I try to be pleasant, I find on rereading I grow
silly and stick the sheet into the fire. And when I attempt to write of the past,
I take so little interest in it that I cannot believe that you or any of the rest will
do so. But on the first dawn of better days (if ever they come) we will try to
give you a history of the past.

Everything has went crooked from the day that we found that confounded
opal, and if it don't turn out something why—confound it! Before that, when
we had a run of bad luck, worked ourselves down with prospecting, and
health so much injuries that we could not work half our time—Providence
would come to our rescue, by giving us good diggings. Since then everything
has went crooked. By the bye while I am grumbling I may as well state one
feature of our "luck" in mining. We have never yet found diggings in which
we could make half-an-ounce and upwards per day but what one or both of
us have been unwell, so that we lost at least one half our time. With scarcely
an exception the rule has been the more work the less pay—and this too, in
face of the fact that we always regarded such things as the Oracle told Ulysses
to regard the shadows on his entrance into Hades—harmless, so long as they
were unregarded.

Saturday, May 26th.

You must excuse the grumbling under the last date—things look brighter
and I feel better. Hosea will probably work some Monday—so you can see the
miraculous power of the Governor's ointment. In despair I threw sluice, tom
and rocker to the deuce, and with pick and pan, took it Spanish fashion, and
have made rising $2 per day ever since—and easy work.[16] I think I can count
on $1 per day as the minimum—that is to say make that <u>every</u> day—generally
over. Hosea will not do much work at present—in fact I do not wish him do
much of anything in which the back is taxed. As for me, I will soon be stout
and hearty—as today is warm and probably the first of summer. Our changes

from winter to spring and from spring to summer, is generally very marked and sudden.

If California can have another dry winter—and I think at any rate, for the mines are much nearer exhausted than is generally supposed—she will be so changed, morally and politically, as to astonish the world. Our last legislature have given us an anti-gambling law—and a chance to vote on the liquor question. Do not be surprised if the people adopt the latter by an overwhelming vote. This section is unanimously in its favor. Within two years California will be the strongest anti slavery state in the union.

It is just as well that we are not at home at present, for if we were we would both go to Kansas. All you want there is a few men who are not afraid to use the revolver, for the Missourians are as a class, cowards and cravens. One Californian would be a match for any "flock" (as the Missouri "crowd" is here called) of 8 or 10. Mean spirited, cowardly, unscrupulous, they will take any advantage that reluctance to use violent means on the part of the free settlers will give them. I don't think that I am very bloodthirsty, but really I could not help wishing that I had been there on the day of the election.

The struggle between Slavery and Freedom has commenced in earnest, and I must say that the tone of your letter gives us more hope than you seem to entertain. If the Friends of Freedom[17] are fearful of failure they are safe. If they are only vigilant now before the election of '56 the wild spirits of opinion in the South will become so unruly as to break through all restraint, and spoil any plans the "old heads" may concoct. The fire-eating spirit fostered so carefully by the great Southern leaders is now grown so powerful that it is uncontrollable and they will find it so. They have raised the Devil, and the Lone Star Association[18] cannot lay him. The great men of the "chivalrous"[19] young men have become persuaded that they have really been most grossly insulted by the stand taken by the North in last fall's election, and have got their heads full with coffee plantations and a black Republic around the Gulf of Mexico that they really are in earnest, and the great mass of them will only be driven to desperation, at the hopelessly of dissolution of the Union, or the terrible risk of a war with England and France on the forcible seizure of the island of Cuba with the chance of "immediate emancipation" on the Atlantic coast. "Whom the gods would" etc.[20] In this section of California the feeling is very positive and decided that Slavery should be confined where it now is, and I think this is the feeling throughout the state. The Southerners here cannot keep their tongues quiet and the consequence is that Northern men with the strong attachment to their respective states which marks the Californians, say many things that go down hard with their hot-blooded brethren. They are told pretty plainly that it would hardly be policy in the South to bring the Canada line down to Mason Dixon's in case of a peaceable disruptive—that an appeal to arms is too absurd to be thought of.

Hosea goes to town tomorrow, and I have written with a bad light, in the smoke and in a hurry—and send what I have written for want of better.

When we shall again see you from present appearances is quite uncertain, and it is folly to set any time at which we are likely to do so. Meanwhile rest assured that it will be as soon as we can, and that our thoughts and feelings are always with and for you. We have shown, I fear, too little interest in the affairs of the family in our letters. But they are indeed but a poor index to our feelings. We often commence letters full and complete as we have the time— paint our hopes and fears—our trials and troubles when, getting homesick, or disgusted with the confounded "luck," which is forever spoiling while we are trying to build, they are thrown aside, so far from completion, that when the time comes when it is necessary to write, we snatch up the pen [and] scrawl off a hasty sheet to let you know that [we] are yet alive. We have so much to say to all of you that the task has grown Herculean, and nine chances in ten when we write we find that we have left unwritten what we most wish to have said. The letters we get from you all, are very, very interesting, though we do seem to value them so lightly by giving such poor goods in exchange. If we only could get to our feet again things would be different. If appearances do not again deceive us, and our present prospects hold out, as moderate as they seem, we will soon get up again, for we [are] very economical in all our expenditures. One great good that will result to California from the present hard times will be the abolition of two-thirds at least of supernumerary go- betweens and idlers that the poor miner is now compelled to support in buy- ing his provisions, etc., from the retail trader. From an examination of the Price Current of San Francisco one would suppose that living here would be not much dearer than it is at home but by the time it is dealt out to the miner it requires economy and care to keep his expenses within $1 per day. We man- ufacture nothing, raise but a small quota of our vegetables, and furnish no employment but gold digging, and if ever a country has been cursed by the fruits of folly, California is that country.

This summer will exhaust the mines in the rivers. Next winter will open well and, for the number of hands employed, will produce a large amount of gold. This will make times easier, revive hope and encourage speculation. Towards spring the supply will fall off, and by the end of the rains sink to comparatively nothing, and there ends the placer mines of California. It will no longer be the leading idea of the state—a happy day it will be! For she will then commence a second growth, solid firm and useful, relying on her strength as a commercial and agricultural state.

Tell the girls that we will try and write soon.

Give our love to all.

Truly and affectionately your son,

E. A. Grosh

ᕯ *Letter from Allen Grosh:*

July 29, 1855

Mt. Sugar Loaf, California

Dear Father,

We have been waiting for this last five or six weeks to hear from you concerning our stone. We received a letter from Uncle Tom Disheroon stating that in New York they had been pronounced water crystals. But as Mr. Lutzinski the jeweler at Mud Springs insisted that they are opals and nothing else, we have waited anxiously to hear from you. Mr. Disheroon's information is comprised within a sentence and is far from satisfactory.

Hosea's health is improving slowly but steadily. He does not work, but by hunting, at which he is quite skillful, he keeps us well supplied with meat, so that though we make but little at mining, our living is both excellent and cheap. Your missive concerning the Tartar of Antimony[21] came, probably, in good time for it began to lose its good effects on Hosea, and bad ones might have ensued. He has since followed mother's advice with certainly some good. Bathing has in California a remarkably good effect. Many of our low "typhus" fevers which give our physicians so much trouble yield readily to a daily bath in pure water.

My general health is improving rapidly, though I am still troubled with biliousness—probably induced by the hard water.

Hosea cooked and I worked in the field during harvest. After two or three day's practice I made a full hand raking and binding.

We wish to correct an impression which both you and mother have received from Mr. Disheroon. Did he not use the term "Too Benevolent"? We think so, as he received the impression from our friend and neighbor Mr. Hoover, or The Governor, as he is generally called, and whose great prominent fault is "benevolence." It is a fault long since corrected in us. But the old man still holds to it as a means of explaining our ill success, as it in no way injures us in his esteem. Like the Deluge to the ancient geologists it serves as a general explanation to everything he did not clearly understand.[22] He thus saved our ability from the fatality attending our fortunes! So far from being extravagant there are but few in this section as economical as we. We live well, as is true, as a matter of economy, but we do it on fifteen or twenty cents per day. Hosea shoots most of our living with the rifle. No, father, had we had ordinary luck we would now be comfortably wealthy. That we did not always pursue the best and wisest course is readily admitted but even most of those mistakes are in part excusable, for our greatest blunders occurred when our health was poorest, and when we were least fitted for thought or action. There is poor consolation to be drawn from the fact that when we were most painstaking we were the least successful.

To cap all, this past winter with health good enough (on my part at least) to manage carefully and work—and I must say that I think we were neither overcautious nor over-bold—we have made the most complete failure since we came into the country. Everything went against us. And yet last fall and winter we did more prospecting than any six persons in the diggings yet we found not one place in which we could make $2 per day for over one week at a time. This spring I have worked week after week for $3, $2, $1, and less per week. We have had places given to us by friends to "break our luck." And in a few days they would completely give out—no matter how well they had paid before. Now, had we run in debt $10, $20, and $50 . . . in prospecting places, it would be plain enough why we found nothing, but we adopted the Chinese maxim to make what we could and be content. Other miners—a few—in this section made nothing and run themselves into debt—but I feel safe in saying had they tried as hard as we did, they would have come out ahead.

But enough of this complaining—it is something I am not used to—I verily believe that it is all for the best. It has already given us a nerve—a sternness—to our characters, we were strangers to before. We stated these facts that you can see what a game of chance this mining is.

We are now about throwing ourselves on fortune for the last time. We will commence Monday throwing up dirt for the coming winter. The place we have picked out is one which from its dry situation has never been thoroughly worked. It has been rich, and we think is so still. We will work it by using the false bottom sluice, during and immediately after each rain. This sluice is so constructed that by a false or second bottom the gold is retained and the dirt washed out by a heavy head of water. By this sluice much of the labor of gold washing is saved, as it requires no further attention than to seeing that the dirt washes through. It is by far the most perfect gold washer in use, and is the only one by which all the gold is saved. Every three days rain will give us four days use of this machine which requires so large a head of water that it is not much used in the dry diggings. When the head gives out we will at once "clean up" what we have washed through, and then prospect and throw up for another rain. We think, from our knowledge of the place that it is safe to count on $1,000 each, from the present time to the first of May. Our success, however, will depend much on the amount of rain falling during the coming rainy season. We will be compelled to throw it up "blind"—that is, without prospecting—as there is no water near. However that makes but little differ-ence. We have been deceived so much the past winter with pan prospects that we have less faith in them than ever. The pay and extent to which a place is worked form better ground for judgment. It paid half an ounce to the hand, last spring when the water dried up. Our district is so secluded, and the min-ers so scattered in it is the only reason that it was not taken up a month ago. But a few of the "Old Heads" knew its value.

The present state of California, though much improved since last winter and spring is truly deplorable, mining being the only business in this part of California. We have finally so far exhausted the field of labor as to be unable to give room at living prices to our population. I was once inclined to free trade but I must say that California has effectually cured me. It seems strange that though wages are reduced to $1 and $2 per day tradesmen and mechan-ics find no encouragement to start in their respective trades. We still draw our clothes, implements, and the greater part of our living from the older states and foreign countries, thus giving away the gold we so much need in mak-ing the transition from a mining to a commercial and agricultural state. This gold digging is very much like drunkenness—we cannot grow sober while the bottle is unbroken.

The rivers are worked this year with a degree of skill and prudence here-tofore unknown. The result has been fewer failures. But the heavy expense attending mining there, renders a failure absolutely ruinous. Months of patient, heavy, laborious work, mostly in the water, and hundreds of dollars of cash or credit are necessary, in most cases, before any retrieval is made. These investments are made, generally, not by capitalists, but, by miners. If he succeeds he goes home, if he "gets his money back," he thanks his stars, and thinks himself lucky; if he loses all disheartened and discouraged he seeks a refuge in the dry diggings to prepare for another winter's work or starts of "North," or "South."

But this year I fear we, the unfortunate will have no place to go. The "South" and the "North" from Kern River to the Cosumnes and from the Cosumnes to the Feather are "worked out." If he goes "Way North" into the Shasta and Trinity country he will find many rich claims with many—too many—men to "work them out." He will find all the mines of California much in the same state of those around him.

Throughout the dry diggings are remnants of large claims and little spots dryly situated, or discovered last winter too late to be worked to advantage and left over for another season, the roughly prospected, only. These are all held very firmly, and few can be bought at a fair price. A few more will be discovered when the first rains give water for prospecting—others by accident during the winter.

The great mass of the miners on the river who are in any way successful will leave for their homes, or turn their attention to something else, for except those that have held on to claims in the dry diggings, few will find footing there. Of the surplus of miners now in the dry diggings those who do not find a profitable place, will also leave—at least those who can—before fall. The result will be that the coming winter will yield a larger amount of gold per man than has been yielded since the opening of the mines.

The "hard times" of the past eight or ten months have taught us the miner's economy, and seven year experience with the constantly decreasing quantities of gold as the average yield of a day's work, prudence and skill. The extremely dry winter of last year, while it prevented constant working, gave opportunities for prospecting, such as we never had before. It gave both time and water. For these reasons we need anticipate but few failures.

Unless the coming rainy season "dries up" as effectually as did the past I think the above calculations may be relied on. If on the other hand we should have an extremely wet season, like that of '49—it will increase the amount largely—for there are probably but few sections in which the miners are not acquainted with patches and spots of ground which would be valuable only in such a season.[23]

The following summer a decreased force of miners will find profitable employment on the rivers to finish up the work of the present season, and then gold washing will sink from being the leading interest of the state to one of secondary importance, where though not as prominent, it will, it is to be hoped be more useful. At least if it prove not a blessing let us hope that with its preeminence will be removed the curse that has breathed the spirit of the Wandering Arab[24] into the souls of our people.

The object I have in view in dwelling so long on the state of California is to point out [to] you another "money crisis." It takes but a short time for the gold taken out of the mines to find its way to San Francisco, and from thence to New York. The knavish speculators and credulous dupes below, will, on the strength of the large shipments of gold, raise once more the cry of good times, and the "inexhaustible" gold fields of California, and by false representations induce another rush from the states. In this they will be aided by several causes worth pointing out.

1st. There is something about gold digging that in itself favors exaggeration and excitement, and there are but few miners but who, after a month's good luck, think it is going to last forever. They are very apt, too, to attribute too much of their success to their skill or industry—their "saba"[25] as we say here. The people in the East will not receive counter testimony from their friends here. The general tenor of information from miners will rather confirm the testimony of those interested in getting up the Rush.

2nd. Our deep placers will continue, probably, for the next 8 or 10 years, to yield as abundantly as they have heretofore. . . . Trask[26] compares them to the coal beds of Pennsylvania, and thinks their yield will be as certain. Though his evidence must be received with caution, yet, from what I can learn, in this he is probably correct. These placers require not only large capital but skill and judgment to work them with a <u>certainty</u> of return. The "large strikes" made here will be cited as proof of the "inexhaustibility" of the gold fields of California, and no mention will be made of the capital, skill and labor expended to reach those deep deposits.

3rd. Quartz mining is reviving all over the state. Many of the mines opened and worked in fifty-one and two, and then causing ruin to all concerned with them, have, under judicious management, and cautious experimenting, proved to be profitable concerns. If followed up judiciously it will become an important interest, at once; but if the inflating hand of speculation touches it, it will blow up, and be thrown back 4 or 5 years longer. It is a beautiful business for the speculators, for I doubt if there is any other in the world more calculated to deceive a person not acquainted with it practically. It was made such a great humbug, and so many were so outrageously bit by it in 1852, that, possibly, the lesson then taught will be remembered now. I fear not, for capitalists are fluttering about the trap like singed-winged moths, anxious, but fearful, and occasionally making a dash, . . . The confounded fools! . . . This quartz business, too, will help swell out the reports which will be sent you.

Prosperous times, dependent in any way upon California, are not to be relied on after the first of next June. An extremely dry winter—and we may have it—can, I think, alone prevent what I have predicted. By the middle of October I think we can tell.

Since we last wrote you we have moved our residence again. We are now about ¼ mile from our last place. We occupy a large, roomy cabin, beautifully situated, with a large plot of ground suitable for a garden. The spring (a very good one) is situated so high above this plot of ground that water can be carried to any part of it. Should we find a <u>good</u> quartz vein, it may probably induce us to stay, in which case, we will at once plant fruit trees, etc., for a home. We have also our notice on the only real good plot of ground for farming in the section. It is situated on the top our old ridge, about half-a-mile from here. It is almost level and has a good soil. But these are only possibilities. We have had so much bad luck here lately, that, California-like we are tired of the place in which it happened. We keep our hold for fear of doing worse.

We <u>think</u> we know where there is a silver mine on the other side of the mountains. Antimony and native silver in small quantities, we <u>know</u> occur on the same stream a mile or two below. The one we found was the black silver ore, if silver it was, in masses large as your fist. We were so hard pressed by poverty, while on the other side that we never tested it. Since our return we have not, for the same cause, been able to get a work on mineralogy. Please inform us if you can, if there is anything resembling that ore? If we ever can get means enough to do so, without sacrifice, we will go over to take another look. We think the vein could be struck without much trouble. Had the stone turned out anything we would have gone over this summer, as in reality we had more hope of finding that than "opals."

Hosea's health did require, very much, a trip into the mountains. By hard work and scraping we managed to raise some $30, but the young man (William Louget,[27] who will work with us the coming winter) who was going with him, was disappointed in his expectations of getting some that was due him. The trip fell through, and they were compelled to abandon it though with great reluctance. The trip required a horse, which went beyond our means.

And now, dear father, having used up all the paper in the house I must stop, though I have considerable more to say.

Kiss Emma's and Letitia's babies, give our love to mother and the girls and Warren and believe me to be,

Truly and affectionately your son,

E. A. Grosh.

✐ *Letter from Hosea Grosh:*

December 10, 1855
Sugar Loaf mines El Dorado County, California
Dear Father,

About three weeks ago we received Malvina's letter, and a week later one from you, announcing your intended removal to Perry, New York and were glad to hear it, having conceived no small prejudice against Fort Plain.[28] We have strong hopes that your new location will suit you much better. From what we can judge, that Fort Plain is decidedly an old fogy place, devoted to all old fogy notions, and naturally opposed to progress; therefore, our joy at your leaving a place so ill suited to you. Tell Malvina that we sympathize strongly with her and feel glad that we have so brave a sister, who can rebuke what she disapproves even though it offends the wrong doers. In this state we have done tolerably well on that same liquor question in the vote for prohibition being but 5000 votes behind; our corrupt cities going as they did dead against it. Bad as the mines are, they are the best (in point of sense and morality) part of the state. We would have written immediately on the receipt of your last, as our consciences reproached us with neglecting you so long again, but were so busy moving, building, and fixing up our little tent house; so as to make it comfortable, that we have not been able to do so until now. Our only excuse for former neglect is, that our life is so monotonous that it gives us but little to write about ourselves, and you have no interest in anyone else here so we had nothing to write and so it was put off from day to day, so excuse us if you can, and we will try again to do better, after while we may succeed.

During August I was kept confined almost entirely to the house by two felons,[29] on the middle finger of the right hand. They caused me considerable fever and pain if I moved about any, but by keeping quiet, I got along pretty well, and almost without pain. Heretofore, I have always taken them in time, and killed them in the bud, but this time having run a rose thorn into the end of the finger, I did not discover it to be a felon until too late; I will keep a better look out in the future. In building the house I was taken with a kind of lumbago in the hips, and stopped from all work for several days. This still bothers me a little, but I will be rid of it in a couple of days I hope. My health is better since the cool weather has come and I hope will, with care, improve during the winter. I have done but little during the summer except keep house and hunt. We are throwing up dirt so that when the water does come we can take the fullest advantage of it. We have had several showers and one day's rain within the last week and hope that our rainy season will so set in strong enough, to give us water, for washing out our dirt. We are ready for it, let it come when it will, and hope we will this winter and spring coming, do our last gold mining. The summer we will spend in the mountains and probably in the fall go into the San Bernardino country near Los Angeles and see how we like settling down there. It has according to best information all the advantages of soil, climate, and health that can be found anywhere. But this is far ahead. We may even go to Kansas before that time. The Kansas question has awakened a strong feeling in this state and if the invasions continue let Missouri look out for the Californians. It will at any rate insure this state to become more of a free state with her politics and with the aid of Know Nothingism enable us to shelve all our old politicians and wire pullers and then let old fogies and Know Nothings look out and keep quiet, for they will have little to do and may lie still where they are put.[30] The people are waking up even here and when they speak California may become a decent state. Our cities, especially San Francisco, are as corrupt as possible and that is the reason that they went for liquor. San Francisco is rapidly going down. Business is dull, the city cannot even repair the streets, and as many of them are built over the water it is really dangerous to go around, a great many accidents occur continually. I expect that this winter there will be a great deal of gold taken out as immense quantities of dirt are thrown up which enables the miners to do at least twice as much when water does come. Besides machinery and its management is probably as perfect as possible or nearly so, for really it seems as if no further improvement can be made, unless someone will invent a digging machine. The winter promises very fair indeed for a wet one. From all the signs I think the roads will be broken up again as in '52, in fact so far, this much resembles that year. We will probably in that case have clear in the spring close on $1000 unless bad health should interfere, which I think it will not to any great extent. The water is decidedly better here than where we were living during the summer and that will make a great difference, we will soon have rain water and that will be still better. We have moved only about half a mile from where we have been living and are but little over that from our old place but the water in almost every spring is different, coming through different rocks. Allen's health is tolerable and I think improving. Since our removal he has worked pretty steadily during the greater part of the summer, for the last two months throwing up dirt, so we have considerable thrown up which we hope will pay well. Our claim is in a ravine on a side hill so that we will only have water during and for a short time after each rain; we have an awning to work under, which will be much better than gum clothing, which generally causes one to sweat too much for comfort, and does not let the perspiration escape; we have therefore determined not to wear it, and take this plan instead. Most of our work will probably be with the rocker except during the heavy rains: We can rock through from 400 to 500 buckets of dirt per day. Everything now seems likely to [move] along smoothly enough though what may come we don't know, and don't trouble ourselves about, being pretty sure of trouble enough for comfort without that.

If I had anything more to say this letter would continue to grow, but I don't think of anything else now.

Tell 'Vine I will write to her in a couple of weeks.

Your affectionate son,
Hosea B. Grosh

1856

Based on handwriting it appears this letter is from Allen Grosh:

February 9, 1856
Hermitage
Dear Father,

I have just come over from the Middle Fork Cosumnes where we are now located for a time. The winter proving a dry one, we determined to go there so as to be making something all the time as the prospect of washing much before the spring rains commence was but slim. We hold on to our claim here and whenever it rains will come and wash. The distance is about six miles. We were about to write before we went but neglected to do so and therefore send you these few lines so you may know what we are about. Our success has been but small, being able to make but about 50 cents per day apiece until yesterday we opened a new place about 300 yards further down the river and in about ⅔ of the day took out $4. It promises well and think it will hold up to prospect at least.

Our healths are pretty good and we expect will continue so as the river water agrees with us much better than that hereabouts.

There has been a big lump taken out by Governor Hoover and company which has created quite an excitement. It is gold and quartz and weighs nearly 12 pounds and is estimated at from $1000 to $2000; certainly no man deserves it better.

We hope to do pretty well when the spring rains come.

We have nothing to say further to say at present but will write more at length soon and close the other as I have to go back again today.

We remain affectionately,
Your sons,
E. A. and H. B. Grosh

Based on handwriting it appears this letter is from Allen Grosh:

March 29, 1856
Mt. Sugar Loaf, California
Dear Father,

We are neither of us exactly well being so much troubled with vertigo as to make it difficult to write a letter. But we wish to let you know of an important idea which we have worked into shape. The great axiom of <u>mechanics</u> has heretofore been <u>that power cannot be created</u>!

That the distance the power moves multiplied by its weight will equal the resistance overcome multiplied by its distance of motion added to the friction. In formula X = distance the resistance or weight moves. (P × D = fr. + (W × X)). Therefore Perpetual Motion was impossible except by entirely overcoming friction in which case each side being equal it might be done. Eubanks[1] mentions a <u>hydraulic ram</u> on the Rhine which with a fall of 3 feet raises two thirds of the water up to a castle 300 feet above instead of four and a half feet as would be the case if the above mentioned axiom or law was correct. The column of 300 feet high would give a pressure of 135 feet per square inch equal to that of steam in ordinary engines and capable of application in the same manner as steam to all high pressured engines. By the direct connection with the ram instead of raising up the water and letting act by a tube the pressure must be the same.

To make a self acting or perpetual motion machine we propose to throw the water used in working the engine into the reservoir at the head of the ram and employ a portion of the power to pump back the waste water to the same place. The vertigo has compelled us to leave and otherwise interfered with us or we would probably have a model at work now. I am going over to town to get some medicine to cure and Governor Hoover thinks that a few days will see us all right. We hope so. We will write again in time for the next mail and will probably repeat the substance of this letter in that to insure it's reaching you.

I forget to mention that the engine worked by water called the pressure engine is identical with the high pressure engine in its construction.

If the above plan succeeds and we can see no reason for its even partial failure we have solved the great problem of a cheap motive power to supersede steam besides inventing perpetual motion. We will make a model as soon as convenient.

We remain your affectionate sons,
E. A. and H. B. Grosh.

Letter from Allen Grosh:

March 31, 1856
Mount Sugar Loaf, California
Dear Father,

Yesterday, Hosea, in going over to town to mail our note of the 29th and get some medicine prescribed by Governor Hoover for our vertigo, received yours of January 29th, and Warren's of February 5th. Today I am so well as to undertake writing this—yesterday my head was in such a whirl I could not have thought of it. Warren mentions that grandfather is afflicted considerably in the same way, and therefore I mention the prescription: 20 grains of quinine in a pint of good brandy. A teaspoon full to be taken about 15 minutes before each meal. The Governor says that it is caused by inaction or torpidity of the nerves of the stomach. Hosea suggests that grandfather's may be what Dr. Behne would call the quinine fever, as the symptoms in our case were almost precisely those of the after effects of quinine. Grandfather, he says, took large quantities of it for fever and ague. In our case it must have been caused by something in the water of the Middle Fork of the Cosumnes River, as most of our neighbors there were affected in the same manner—besides it is a long time since we took any quinine. We have suffered from it more or less for 5 or 6 weeks. We came back here eight or ten days ago on account of it and last week—it being cloudy and threatening—suffered severely. The Governor thinks it nothing serious. Otherwise our general health has much improved. We just paid expenses on the Cosumnes.

Your letter gave us much to rejoice and much to grieve at. You are at least settled in a neighborhood composed of better elements than old fogy Fort Plain, and we do hope most sincerely that the people of your new home are what you expected. But our family and their friends—how they are scattering! Warren alludes to it in his letter. Is there no help for it? Can we not at any rate strike on one common interest and all move in one direction?

It is, we know, a subject on which you feel deeply, and though we have said little to you about it we beg to assure you it has always been in our thoughts. Our hope has been that Providence would yet bless our exertions in California and place the means in our hands—and we cannot yet bring our minds to doubt it. All the family have been trained in the bitter school—adversity, and but little—comparatively—is needed for a start.

In the machine described in our note of Saturday we think we have the means. We have examined it as carefully as we possibly could, and with the utmost skepticism for the whole thing is so simple as to excite astonishment that it never was observed before. While on the river, we had not the means of making a model—since our return the vertigo has prevented all thought of it.

But the machinery is so well known to mechanics and their powers so accurately determined that there can hardly be a doubt as to its success.

It is well known that the steam engine can be worked as well by water as by steam, and that the same power can be obtained by the former as by the latter, by having the column of water of a sufficient height to gain the requisite pressure. Now suppose you connect a ram to a steam engine of 5 horsepower in place of a boiler and have the ram of sufficient size and power to fill the cylinders as rapidly and with the same pressure as steam; will not the result be the same—minus the power necessary to pump back the waste water? We cannot doubt it. If this is correct what cheaper, better or safer motive power is wanted? A patent to cover its application to the pressure engine, and all other water powers must be a fortune of the greatest magnitude! And cannot such a patent including the self moving principle be obtained? Oh that we were home with you, if but for a few weeks! A very little trifle of time and money would settle the whole question—while here it is an entirely different thing. Pressed by poverty, and away from the material for constructing the model, we can only put our spare labor on it and build it as best we can. We hope before the next mail leaves to have one working.

That there will be difficulties to overcome to make it practicable, we have no doubt. Our greatest obstacle is our ignorance of the ram, and the impossibility of supplying ourselves with information, books, etc.

We have kept the whole thing a profound secret, no one knowing anything of it except yourself. We will probably say nothing until we hear from you. You are at liberty to do as you think best. We should like to have Reuben's opinion, if you think best.

Failing in this, we have another hope. Ever since our return from Utah we have been trying to get a couple of hundred dollars together, for the purpose [of] giving a careful examination for a silver mine in Gold Canon and though we have worked, economized, and starved, so far it has been in vain. To be sure we may be mistaken for we made no test, but we had a catalog of minerals with us, and we found an ore which answered the description of carbonate of silver precisely. Now native silver is found in Gold Canon, but tarnished by—probably—the sulfuric acid in the water. It resembles thin sheet-lead broken very fine—and lead the miners suppose it to be. The ore we found at the forks of the canon—a large quartz vein,—at least boulders from a vein close by here shows itself. Now the carbonate of silver is too soft a rock to wash many yards, and we found it in masses larger than your fist, a few weeks work will, we think determine the matter. Other ore of silver we think we have found in the canon, and a rock called black rock—very abundant—we think contains silver.

Tuesday evening Mar. 31, '56

If we cannot get sufficient means to make the examination this summer, we will have to wait until next. We have but little fear of anyone else finding it, as all traces of it have been destroyed. A few weeks labor will determine the whole matter.

This thing is certain: if we ever have the means placed in our hands it shall be devoted to gathering our family together if it is possible.

As to our coming home at present it is utterly impossible. We not only have not the means but are behind about $100—and that, too, in the face of the most desperate struggle to keep even. Fifty dollars we could lay out in clothing and necessaries now, and not be more than comfortable. For the year past we have made our own clothes, and during the same time we have bought two pair of shoes! As to boots they were not to be thought of! Our living has been of the simplest kind, though generally we have had an eye to health. Meat, shot by ourselves, has been about our only luxury. Now game is scarce, and we don't attempt to shoot any. Now why is all this? Any old Californian can tell you. We have "got a streak" of bad luck, and of course everything goes crooked. Besides, we both verily believe that could we have got $100 ahead, we would have struck our silver vein and done a handsome thing by it. This cannot last forever—it must turn and soon, and if our "good" equals our "bad streak" look out for the 100 pound chunks!! But indeed—indeed—we have had a hard time of it—and must not be scolded too hard for neglecting our correspondence. Tell 'Vine so—God bless her!

We got to work today for the first time since coming back from the Cosumnes River and made $2.50, short day's work—and promises of better ahead. The places we are working in offer a fair chance for coarse gold.

It has been trying to rain now for more than a week—but so far the water has only been "started" in the small ravines. The claim on which we have most relied, we have only worked on for about a day—ground sluicing. We think that three day's work there would pretty nearly square us, if not more. So far the rains this winter have proved utter failures—bad luck to them.

The Governor found a lump on his claim this winter worth nearing $1300. We will probably give our quartz vein an examination soon. The company will hold a meeting as soon as the rains are over.

Give our love to all

Affectionately your sons,
E. A. and H. B. Grosh

P.S. Our vertigo is entirely gone.

✐ Letter from Allen Grosh:

April 17, 1856
Mount Sugar Loaf, California
Dear Father,

We proposed writing more fully by this mail than we do; but our health continues poor. It takes very long to recover fully from even a slight fever in California. Since the last letter we have worked pretty steady—though not hard—and with better success than usual, for we have made a comfortable living, besides supplying ourselves with many things we have long wanted. We have also secured a claim, on which a year's work can half be thrown away. We think we can count on it with tolerable certainty—anybody else might and would, with absolute certainty.

It is impossible to make a model of our machine, as we are now silver miners . . . and have accordingly laid the whole machine by until such a time when we may have may arrive. We regret this necessity exceedingly but so it must be.

The difficulties in Kansas excite considerable feeling in our neighborhood, and if we mistake not throughout the state. Our means of judging are very poor, but we feel confident that California will stand firm for freedom in the coming struggle. The appointment of the commission for examining into the Kansas question has given us great hope and satisfaction. Another sign of hope, that of Dunn and company[2] taking the lead in the matter. They begin to find out that the people of the North will bear with no further trifling in the matter. Our neighborhood, the Sugar Loaf, is strongly Free Soil. At the last Presidential election, five Democratic votes, with "For Presidential Electors" a blank gave our strength at our precinct. This year on an increased vote, the Republican majority is certain. Will not this be the case throughout the mines? When you recollect the character of the miners, formed by the business they follow—self reliant, liberty-loving, thinking, hard working, and manly—I think it quite probable. Besides, our mining population have by intercourse with all phases of character, become shrewd observers, and hard to be deceived—politicians cannot lead them.

One subject we wish to mention, as giving us peculiar satisfaction, namely, grandfather's autobiography and we anticipate much pleasure and instruction in its perusal, if we are ever able to get back to the Atlantic side of the continent again. How we forgot to mention it before I cannot say, but each time after mail our letters ever since you first informed us of it, we remembered with regret our negligence.

As we will not be able to write to Warren before next mail, please tell him if you write that Sacramento is our capitol de facto, Vallejo near Benicia de jure.

Our legislature, the people, and the Supreme Court, have each been at log-gerheads about it with the other. The Supreme Court for some time insisted that San Jose was the true and lawful capitol. Sacramento is the choice of the legislature.[3]

My head is somewhat feverish this afternoon so that I must stop. Hosea goes this evening to Mud Springs.

Dearest love to all.

Truly and affectionately,
Your son, E. A. Grosh

～ *Letter from Hosea Grosh:*

May 15, 1856
Sugar Loaf, El Dorado County, California
Dear Warren,

We received your letter near six weeks ago, but being neither of us, well, and not feeling like writing more than we <u>must</u>, have left answering it to the present. We went about the first of February to the Middle Cosumnes and while there were attacked with a kind of vertigo fever which compelled us to leave in March. It hung on until we commenced taking medicine for it. The first place we stopped at, after coming back to this neighborhood had rather bad water, so we moved again in April to another spring, where the water was good. Here we still remain. Our health is tolerably good now and we work steadily though not hard, making our expenses easily. We have a good claim on which we have water, enough for the rock all the year, and can make expenses all the time, and when water comes will do very well. We are to have a ditch in to about a mile from here soon, but it will hardly reach us before next winter, when we will have water, rain or no rain. Our weather has been precisely the reverse of yours, not only in regard to temperature, which is a matter of course, but while you have had snow, we have had dry weather, the driest winter since we have been here, I doubt whether it has rain more than 6 or 8 inches in this neighborhood, during the whole winter, in the northern part of the state they always have much more than we do.

We were pleased to hear that you were so well satisfied and getting along so well. In answer to your question: Our capital is Sacramento. It was formerly San Jose, but General Vallejo's[4] promising to erect the necessary buildings, at a place bearing his name, near Benicia, the people by vote removed it there; but on the legislature meeting there the necessary arrangements had not been made; whereupon, they removed to Sacramento, without consulting the peo-ple. The Supreme Court for a time insisted that San Jose was the capital, and held their sessions there, but finally removed also, to Sacramento, where the

capital will probably remain. School teaching will not, I fear, be either a pleasant or profitable business in this state, until things settle down, which does
appear likely to be, before the mines are disposed of in some way. We cannot
say how long we will remain "old batches,"⁵ present appearances would fix
on long term, but no one can foresee or control his fate; Alla illa Alla!⁶ We
will probably start for home about a year from date, perhaps sooner should
we be fortunate enough to have the necessary funds to spare. It seems long to
us since we left home very much longer than it really is; for as we look back
the number, variety, interest, and importance of events, is what we see our
landmarks to judge of the distance we have travelled, rather than the distance
between them. Since we have left home events, many, various, and exciting
have followed in quick succession making as it were many eras in our lives.
We are no better in money matters than when we landed at San Francisco,
in fact not so well off, but then if we could only find some good investment
for our experience we would soon rise in the world and no mistake. The only
thing we have gained from it so far, is some little philosophy in enduring bad
luck, which we take as easily as anybody. We do what we think best, and don't
worry ourselves about what we can't help, even if things don't go to please us.
If that isn't good philosophy I am no judge.

In your letter as well as father's we see that our family seems likely to scat-
ter far over the west. We are sorry it is so, but have hopes that may all be gathered together again. God grant it!

Politics are just beginning to be stirred up but we expect a pretty warm
campaign the coming summer and fall. We do not know how you stand in politics but hope that we agree. For us what we have seen here of men from all
sections of the union has given us an excellent opportunity to compare those
from free and those from slave states with each other and feel satisfied that the
"Peculiar Institution"⁷ does not improve the white man whatever it may do
for the black. And it has fixed our resolves to do what we can to prevent the
spread of the monster evil. In fact had we had the means last summer I feel
certain we should now be in Kansas. In regard to the Missourians, a word or
two may not be out of place, as we have had the same race, to contend with
in this state. There are of course good, honest, brave men from Missouri, even
among the rough borderers, such as Kit Carson,⁸ and other mountaineers, and
many quiet, peaceable ones; but the majority, of the lower class, are ignorant, cruel, overbearing and cowardly, and these traits have made the name a
reproach in this state. Take our worst of those who go in gangs, and abuse and
insult, all who are weaker than they, and raise them in the back woods with
the same principles and you have what we mean by "Missourians." It is about
the worst insult you can offer, to call a man one. One who has some ability
and will drink with them, mingle them, and gain their confidence by acting

worse than themselves, can do what he will with them so long as the danger is not too great. Having no generosity or magnanimity themselves they cannot appreciate such qualities in others. This picture may seem overdrawn but I describe them as we found them here and from what I see of their doings in Kansas I judge they are the same there. In this state one thing is certain the people are sick of both the Democratic and Know Nothing parties. They have been tried and found wanting and we will be disappointed if California does not elect a full Republican ticket.

We are now in our most beautiful season, flowers carpeting the hills in such variety, and profusion, as I never conceived of at home. From the hill, or rather ridge, on the side of which we live, can be seen one the most extensive and beautiful views that can be seen anywhere. When the air is clear, we can see the whole of the San Joaquin, and a great part of the Sacramento valleys, together with the coast, and Contra Costa Mountains, at their nearest points over 60 miles distant. We can see the mountains on Tulare Lake in the south, and the Trinity Mountains in the north, you may from these data form some idea of its extent. If you look on the map, we are about 5 miles west of a line through Placerville, running north and south, and two east of one through Coloma, and about 16 miles south of south fork of South Fork of American River, and within a mile of the Cosumnes. Measure from this spot, if you can find it, the points I have mentioned, and you will know, better than I can tell you, how far we can see. In addition to this when the air is right clear we see it all plainly and distinctly. The beauty I cannot pretend to give you an idea of. We spend our time tolerably pleasantly, working in the morning and evenings so as to avoid the greatest heat of the day. Generally, I leave Allen working and hunt for an hour or two and in this way get our meat, so we get along; waiting for "something to turn up", as Micawber says.[9] In case we do not get home this summer we will probably spend a couple of weeks the other side of the Sierra looking for a silver mine we have spotted there, as well as to recruit ourselves, during the hottest weather.

We now bid you good bye, and remain your

Affectionate brothers,
E. A. and H. B. Grosh

   Letter from Hosea Grosh:

August 2, 1856
Mount Sugar Loaf Mines,
Dear Father,

We have not heard from Mr. Disharoon for a long time, our last letter remaining unanswered. We inquired of his business agent in Mud Springs, who informed us, that though he (the agent) had written several times, in regard to the sale of Mr. Disharoon's ranch and other matters of some importance, he had not received any word from him for six or eight months. Neither he nor we know anything of him therefore and feel considerable uneasiness in regard to him. His honesty is without reproach here though it has been pretty severely tried; so that I cannot think that he would be dishonest at home. We regret your loss but feel sure that there cannot be anything in which he can be to blame in regard to it.

It has been suggested that perhaps his infirmities have affected his mind and his relatives feeling desirous to hide it have neglected his correspondence, or that domestic trouble may have come upon him and produce such an effect. When he left here he was somewhat short of money, having remitted freely home, and failed to collect what he had calculated on getting in, because, of the hardness of the trials.

Allen commenced a model of our "motor" today and is now working with axe and saw at the butt end [of] an obstinate log to snake it out of it. In our next we think that by the next mail we will be able to inform you of our success. It would have been started long ago but we could not think of any way to do it, with the means in our possession, until within a couple of days.

We remain your affectionate sons,
E. A. and H. B. Grosh

✐ *Letter from Allen Grosh:*

August 16, 1856
Sugar Loaf, California
Dear Father,

By the last mail, though not in time to answer, we received yours of June 16th, enclosing grandfather's note and Uncle Franklin's letter. And sad news they brought us! From what you had written concerning Aunt Mary's health we were led to expect that the blow could not be much longer delayed. It has fallen with a two-fold weight and may God in His mercy deal tenderly with the stricken ones. Franklin feels his double loss severely, and oh, how severe the loss must be! We sympathize with him and his young-helpless family— would we could do more—console them. God help him! God help them!

Grandfather's painfully traced characters show how hard the hand of time rests on his gray head. How time passes! Seven years and a half have wrought strange change in all of us no doubt. For one thing we often pray. May Death do no more work amongst us, until we shall have met again. O God grant it!

We have but little to write concerning ourselves. The summer has been excessively dry. This has caused us some inconvenience and a great deal of extra labor keeping us for a month back until a few days ago, busy digging tanks to gather the water over night. During the close of last week and the fore part of this we were so short of water as to be able to wash only 30 or 40 buckets of dirt per day. With ten buckets of water we wash twenty of dirt! We finally gave up tank-digging in despair. The weather suddenly became cool, and the springs ran more freely, and yesterday we washed 70 or 80 buckets,—today 100. At that rate, by mining <u>almost</u> naked, we can make a living. Fortunately we live in the woods and have no women—except Mexican—within two or three miles of us. Altogether we have done much better than we did last summer—we have had enough to eat. Our claim is a good one, and we will <u>certainly</u> be able to make a little next winter.

Our <u>motor</u> bothers us considerably. We have a model under way but for the want of two or three dollars, which our stomachs cannot spare we are at a standstill. By next mail we think we will have it working. There can be no question as to our success. It <u>will</u> work. We wish to substitute a slide valve in place of the ball, at the outlet, to be worked by a small cylinder connected with the air chamber—also pump back the waste water by the same means. It must and will supply the place of steam as a motive however, and we are constantly worrying ourselves with the fear that someone else will hit on the same idea. Preserve our letters concerning it. Hosea's was sent without an envelope so that it might receive the Postmark of Mud Springs, and serve as evidence if necessary. O this bitter poverty! For three years now we have been steeped in it to the very lips. Nothing prospers in our hands—yet we are patient and industrious. We sometimes think that Providence keeps us back until our time come—else why this unaccountable, never ending streak of bad luck? For this past two years we have been too poor to get out of mining, but have been driven on like arrastra[10] horses, blind-folded. Never mind the "Motor" finished, we have something else on hand that will make you all open your eyes. Look out!

Give our warmest love to all. My pen is so mean that it is about as hard work writing with it, as it is nowadays to make "grub" a-mining.[11]

Truly and affectionately your sons,
E. A. and H. B. Grosh.

❧ *Letter from Hosea Grosh:*

September 2, 1856
Sugar Loaf, El Dorado County, California
Dear Father,

We received yours of July 30th, day before yesterday and were pleased to hear such good news domestic and political; especially that Uncle Frank and his son had, in some measure, recovered their health and spirits, after their heavy affliction. Their blow was a terrible one, God grant that there be no more taken from our circle until we all meet again. With us everything has gone and is going as usual, slowly indeed. We have been very short of water, so much so, as to preclude our washing more than four or five hours per day, and now we seem to have worked out all the ground, that pays worthwhile, near water; we have now to wash poor dirt or carry some distance; we prefer the former at present in hopes of finding something in it that will pay us, and because carrying is harder work than any other. In a month or so we will probably have water where the good dirt lies when we will do better than we have all summer. We have also been much bothered trying to build a ram for "motor." I fear we will be obliged to leave it rest until we get a little means. We have some hopes of selling our claims for $200 and if we can, will do so, as that will give us what we want. We know another creek on which we can throw up dirt that will enable us to get as much more in a week or two after water comes. If we do not sell and should be fortunate enough to get a wet winter and are able to work most of the time we will do much better, but wet or dry we can come out in the spring with about that much ($400) but are anxious to get on with the motor as soon as possible.

Just at present I hunt and Allen and Mr. Louget, a young man working with us at present; work and prospect, doing as well by this division of labor as we can in any other way.

Our leisure time we fill up with politics. I have written out one speech which only wants copying and a few additions to be complete, and am preparing another. Allen has commenced one also so that if called upon at any meeting we will be prepared; meantime we talk to everyone we meet. The people generally are not yet awake upon the subject; as usual our canvass will be a short one though I think pretty warm. The Fillmorites[12] made a stir a couple of weeks ago but it has all settled down again. Their tactics were . . . on the administration and trying to save the Union.[13] The Democrats are very quiet; apparently they think the less they say, the better. Their only hope here is that the opposition vote may divide between Fillmore and Frémont, and that they will be able to ruin Buchanan between them and thus carry the state. I see by the papers that the two factions of that party are fighting as usual in San Francisco and Sacramento. We are in a very quiet section and hear but little from the rest of the state or even county except through the papers. In our own neighborhood we are strong for Frémont and think there will be few to vote for the others except southerners and some of them go with us. The question is not well understood as yet and consequently we find many, opposed to slavery extension, favoring Fillmore as was the case several months ago in the Atlantic states. As far as we have seen just as soon as spoken to and the question explained they drop him like a hot potato. There is little doubt that it will be clearly understood throughout the mines before election. It will probably take work, to ensure this state but it will be the fault of the Republicans, if we do not carry it by a heavy majority. We do what we can, but our poverty disables us from doing what we would. We will try to see that for two or three miles in every direction around us the question shall be understood at least, and somewhat discussed before election. If others will do as much we will give Frémont an immense majority, for our people are at heart freesoil but hard to wake up. We should have liked to have subscribed for the *Campaign Tribune,* but were not able to do so, but get by every steamer one or other of the editions of that paper which keeps us posted; it is hardly worthwhile for you to send it to us as it will not reach us before election. The Vigilance committee[14] have closed their labors, for a time, and we fervently hope, forever. If they have succeeded in arousing the people to the necessity of breaking up the rule of wireworking,[15] ballot box stuffing, gambling politicians, they will have done much more good than harm; I rather think they have, but we will see. You will see by the proceedings of the Republican State Convention that they are resolved to have no more "Herberts" sent from this state to Congress. It is time that such men as Herbert, Weller, Gwin, McDougal, etc. were left to rot in private life and not fill the nostrils of the whole people with their stench. I have heard one or two try to excuse Brooks but not one who has dared to say a word in favor of Herbert.[16] The universal thought here is that he ought at least to have been sent to prison for several years, if not have been hung. I don't believe a single delegate will dare to vote for his renomination in the Democratic Convention. I doubt whether he will dare come back here.

We will send you with this one of our papers from which you can gather the feeling of the state better than we can tell you in regard to politics and other matters. As you will see the prominent feature in our news is the great number of highway robberies. The condition of things in this state is deplorable; Through a great portion, he is doing well, who can get a good living; in our own, we are nearly the only ones paying expenses. Where they do better the robber appears to take it from them. You need not worry for us, we are in a neighborhood of old settlers of general good character and so poor as not to induce the robbers to come this way. No one here is expected to have enough to make it worthwhile to steal from him. If we do not get a wet winter this year it will be a question how folks will get along. They have this year even obliged to leave the river in some places for lack of water to work. Next year if this winter is dry too they will have a dry time generally. We will probably do our last mining this winter at any rate.

We remain your affectionate sons,

E. A. and H. B. Grosh.

Letter from Allen Grosh:

September 18, 1856
Mount Sugar Loaf, California
Dear Father,

We will start this afternoon across the mountains to look after our Silver Mine. We may be gone four weeks—not longer. We have associated ourselves with five of our friends who furnish the outfit, so that it costs us nothing. On our return—with improved health, we expect—we will stump our immediate neighborhood for Frémont and Dayton.[17] The canvass in this part of the mines will be short but very hot. We have done considerable in spreading light on the subject and every day brightens one prospect. The trip somewhat interferes with our political plans but it is forced on us by the danger of someone getting the mine ahead of us. I fully believe Frémont will carry California.

It seems to both of us that out of the elements now gathering around Frémont and Dayton, a harmonious, consistent, democratic party can be built which for strength will equal the old Republican and Democratic parties, and we would harmonize the principles thus:

The Slave question. The Wilmot Proviso[18] for the territories. The secession of the District of Columbia back to Maryland. The power of Congress to interdict the interstate slave trade, etc. to be used only for the benefit of our friends in the South. If the Republicans of the South can conquer the oligarchy without national aid, so much the better. If not Congress must aid them as she constitutionally can.

The Tariff:

I take the Seward Whigs and Barnburner Democrats[19] of New York as the type of the members of those two old parties now sustaining Frémont. The Whigs insist on protection—the Democrats aim at ultimate Free Trade—while they would oppose a sudden change that is their aim. Let sufficient protection be given, and throw off on to a free list. Greeley[20] has pointed out this mode of harmonizing indirectly.

On the National Laws:

As a class the Know Nothings sustaining Frémont are moderate. They establish the principles on which our present national laws rest, and correct their abuses—a man who will not make a good citizen in five years will not in twenty five.

What think you of it?

We enjoy usual health. Our warmest love to all. We will write immediately on our return.

Affectionately your son,
E. A. Grosh

∾ *Based on handwriting it appears this letter is from Allen Grosh:*

November 3, 1856
Sugar Loaf,
Dear Father,

We got home 3 days ago, and had as far as the weather was concerned, a very stormy time of it. The snow on the mountains detained us for a week or so. We were very warmly received by our old friends and acquaintances in Carson [Valley], which took the edge off the cold winds and frosts considerable.

We found two veins of silver at the forks of Gold Canon—iron ore containing lead and silver—very difficult to determine. We are very busy now working at it, and will not be able to give the result probably until next mail. That and the election, as you may judge, keeps us pretty busy.

One of these veins is a perfect monster—the other is undetermined. The carbonate of silver, of which we have heretofore made mention, may have came from either of these two veins or a third, the ground for prospecting for which we have secured. We will spend this winter in Carson. By next mail we will give you the result of our assays, and a full account of Carson and adjacent valleys, and the fruits of Mormonism and squatter sovereignty, as developed there.

We received yours of Aug. 26th on our return, and were right well pleased with the news. The changes which have taken place during our absence in the political world are such as to give us strong hopes that Frémont will carry the state. The time is so short and we are kept so busy stirring up our own immediate neighborhood that we cannot look beyond. Again we say everything looks very hopeful. Health very good. Love to all.

Truly and affectionately true sons,
E. A. and H. B. Grosh

∾ *Letter from Allen Grosh:*

November 22, 1856
Sugar Loaf, El Dorado County, California
Dear Father,

It is now raining, but tomorrow it will be clear again, in all probability. If so we will start Monday across the mountains as we can expect two weeks of fair weather. Should the present rain continue it will delay us three or four days.

We did not write by the last mail as we intended, but will try to give you in this a fuller account of our reasons for spending this winter in Utah. The vein we discovered, we took up in the name of the company which fitted us out for the trip. We at the same time discovered traces of another, and richer vein we think, though smaller, in the immediate vicinity. This we did not take up for that company for want of funds and time. We could have done so, making our expenses by gold washing the meanwhile, but we did not wish to work for others and pay our own expenses. We therefore left it secured to ourselves under a prospecting notice. A few days work will enable us to strike the vein in situ, and a few weeks work we are almost sure will enable us to trace the vein down the face of the hill on which it is situated, and strike it at the bottom of the hill. At this point we expect to find right rich rock, as the evidence afforded by the fragments off the vein on the hill, clearly indicates that it grows richer as we descend, and for other reasons to follow. The ore of this, and also the other, vein is the black or magnetic oxide of iron, containing silver and lead, a very common form for silver if we can rely on the information which we have received of the silver mines of Mexico, from various sources. This rock seldom pays enough for working at the outcropping of the vein, but in our veins it not only promises to pay but pay well. It has been pronounced by one who ought to—and we think does—know to be "the handsomest silver ore he ever saw" i.e. not the richest but the most promising. As to the true value of this top rock, we can give you no positive information further than this: it is as you are probably aware, a very difficult ore to melt in the ordinary way. By a fortunate series of blunders we got a result, several times, by the blow pipe—(a reed with a pin hole in it) and a tallow candle. We touched the rock with nitric acid, to see if the acid would affect it. The tallow, probably, also help us in our ignorance. These two results placed the rock above $50 per ton,[21] at the lowest, and, judging from the look of the fragment of ore acted upon (we did not pulverize it), we have reason to believe near all was not got out. This though was richer rock than we afterwards tried. We built a furnace, but in trying another variety of the same rock we split our only crucible— and had afterwards to make them out of the same material with the furnace— a substance very closely resembling the Bath Fire Brick[22]—which, though it would stand the fire well enough, would split on the addition of any kind of flux we could use. We obtained but one result for all our pains and that partial and unsatisfactory. It was but partially melted, and part might have been lost by the splitting of the crucible—our estimate supposed all the silver extracted (which from the look of the rock remaining unmelted could hardly have been) and no loss from the splitting of the crucible. It ran over $80 and probably $100 per ton.

Though we brought some rock back with us, difficulties, from the nature of the rock and our ignorance, prevented our reaching a definite conclusion. The only acids we could get were not only very weak, but impure—so much so as to interfere sadly with the precipitation. After four or five days [of] patient perseverance Hosea got a partial result. The ore we picked out as the poorest,—and we do not think in this we made a mistake, though of course may—it ran over $50 per ton. Now, we could do much better, having both books and acids, if we thought it necessary, but it hardly is, and we have only spoken of it thus fully that you may know on what we base our opinion as to its value. The opinions passed upon it by silver miners places it much higher than the figures we have given. One who is said to be in every way qualified to pass an opinion, speaks of the rock we brought over—top rock—in the highest terms. The cool weather and want of time has thus far prevented him from making an assay of some of it. His mode is one used in the hot climate of Mexico, and takes some days together with a bright sun.

After having given you this account of the iron ore, we will give you our reasons for hoping to strike a richer ore at the base of the hill.

While at Gold Canon in 1853 we had with us a small catalog of minerals by Maw.[23] It was designed principally for cabinet collections, and consequently confined itself mostly to rare and curious minerals. It was published before 1820. We found near the forks of the Canon on the left fork a heavy—very heavy—black rock—resembling somewhat in color black lead. We followed with this rock a description given in the catalogue under the head of "black, or sooty, or scoriaceous ore of silver or carbonate of silver."[24] We have failed to get Dana's mineralogy, but Comstock's "carbonate of silver" is not the same thing.[25] We think it was carburett of silver, or carburet of iron (black lead) and silver. Mr. Maw is evidently a better mineralogist than chemist as in every case throughout his catalogue in speaking of nitric acid he calls it "nitrous acid." At that time the nomenclature of chemistry as now in use was comparatively new, and, probably the substitution of carbonate for carbunet would not be a whit more out of the way than the confusing the terms nitric and nitrous. In partial conformation of this idea the Spaniards all say that the "mark" of the iron ore—which is black—is caused by black lead. I will here remark that another variety—with which we melted our crucible—and which marks red, is said to contain platina[26] as well as silver.

This soft black rock, is what we went after, and expecting to find it we took nothing to test silver with except a little ordinary nitric acid. Where we found it is about 100 yards below our vein. It occurred in considerable quantities. We found on our last trip that the creek had been worked for gold, and all our landmarks completely destroyed, and we worked for two weeks without finding a particle of it. We thought then and do now, that one week's more work would have enabled us to find it. We secured the ground for future prospecting, and turned our attention to what we had discovered. This rock is heavier than iron and if it is what we have every reason to suppose is an 85 percent ore. An acquaintance worked there last winter and says that he found from five to ten pounds at least per day. Now, this rock either comes from our vein or another close by where we found it. It is very soft and easy destroyed, probably not enduring a single heavy frost.

Again, Old Frank, the one who sent us over the mountains in 1853,[27] found in the same neighborhood close by our vein two ores of silver, one of which he recognized in the richest rock we found—the other was a "greenish brownish powder"—probably the <u>chloride</u>. He was in company with Mexican silver miners, and out of both ores the silver was extracted—the one by means of lead, a Mexican process, and the other by quicksilver. A company from San Francisco started this fall on a hunt for silver on information obtained from some of that party—but ran out of their course 50 or 60 miles.

We have hopes, almost amounting to certainty of veins—or what is more probably suite of veins—crossing the Canon at two other points, which can be struck and secured.

We have information of a mammoth vein of copper—copper pyrites—which we think reliable, 25 or 30 miles north of the Canon. A valuable copper mine is now being worked or rather opened in Hope Valley[28]—this vein contains considerable silver. The one we refer to probably does the same. Copper mining is probably the <u>safest</u> mining that can be followed. A copper vein rich enough to pay at the outset generally grows richer.

By going over the mountains now, we will have the whole winter to ourselves without interference from any quarter—next spring there is bound to be a "rush" to Carson Valley—for we do assure you that our discoveries and the little rock we brought back has created considerable excitement, wherever it is known. This winter will enable us to secure all that we shall want if things are as we expect—if it should turn out a failure, we can come back in February. By the sale of our provisions and goods we can nearly square ourselves as far as our expenses go, at that time, and lose only our time—say one month's gold washing here. We consider it our duty to go.

Now what we wish to do is this: form a company of our family and their immediate friends, say 15 or 20 persons—appoint us your agents to take up, to hold, to buy and to sell, and prospect, and work silver and other mines in Utah territory and in consideration of our services as such agents we are to receive one-half of each member's share or claim in addition to our own and the discoverer's claim. This will give us a large controlling interest out of which we can easily give to any who will come out to help us. In the above list include only the family and your immediate friends—for our own friends we can help them if so inclined, if we succeed. By Utah law married women come under common law, and can hold most property only through their husbands. By the rules of the California Quartz Miner's Convention of 1851,[29] which we have adopted, the claim is 50 yards in length of the vein to each individual with one claim to the discoverer. By dividing each claim into a certain number of shares you can make an equitable division, reserving to yourself an interest of at least 2 or 3 claims. The formation of this company, and all connected with it we must leave to you, per necessity, and we feel so much confidence in your ability and wisdom that we hesitate to make suggestions concerning its formation. But one thing we will suggest. By drawing up an agreement, specifying the amount of stock each one is to hold, and having all the members sign it, we think the company could be formed with less labor to you than in any other way. Let the agreement specify the object and everything else then send it to all concerned for their signature and approval. Have a duplicate drawn and signed and send it to us. Make the young folks do the work if much is to be done, only caution them that nothing is <u>sure</u> in mining until you have it.

By February we will probably have either our certain fortune, or make a complete failure. Things look very bright and promising—but experience has taught us that all things connected with mining for the precious metals are very precarious, and that caution, care and prudence are not only necessary for success but often prevent a failure from becoming disastrous. We do not say this to you but to the young folks. Tell all not to build on anything connected with this business. It is <u>not</u> certain.

The unsettled state of our family at the present time makes us very hopeful that in case of our succeeding many of its members will join with us next summer. Uncles Moses and Franklin would be well suited to Utah, and the Mormons would keep their hands full! We think they would both like the climate and country. Then there is Cyrus and Warren. We are afraid John could not be coaxed so far away from Utica.

Today is tolerable clear, though more snow has fallen in the mountains than we counted upon. It will delay us a day or two.

The climate of Carson Valley is probably as healthy as any in the world. The winter we passed there was about as cold during December and January as central New York though not near so disagreeable. It was bright, clear, cold, and bracing. February was delightful. March damp winds and windy every day from 2 or 3 o'clock p.m. until evening—the forenoon clear and warm. April, every other afternoon windy and cloudy—not very disagreeable unless when working in the water, at gold mining—in fact the same may be said for March—it is not near so disagreeable as the east winds of New York. In May you are troubled every third or fourth afternoon with cloudy windy weather. Through June, about once-a-week. July hot, more pleasant, September pleasant for part—latter half windy but not disagreeable. Fall, generally delightful. As to rains you can have them every month in the year or go where they hardly ever come, all within 10 or 15 miles. All the old settlers represent the winter we spent there as the coldest and most disagreeable within their memory—we are inclined to think so—but the character of the Carson Valley people is not very high for veracity—probably the climate is so exhilarating!

We think that if the east slope of the Sierra Nevada was peopled by the right stock—not refuse of California and Mormons—a great race would spring up there. It is so different from California! Instead of the listless inertness of the latter you have a free impulsive energy Californians know nothing about. Once started right and it is bound to produce a race of great workers. As to the natural resources of the country, they are not, we think even touched. The whole country is cut up by mountains, steep rugged and bare, between which are numerous valleys, rich, beautiful and level. The mountains are rich in iron, copper, silver, gold and quicksilver. The distinction between mineral and agricultural lands is very marked and distinct, which will have a very benefi-cial effect on the settlement of the country. A great deal of their "desert" land will probably turn out first class wheat land, as it all seems to be delta and only wants water to make it productive.

We have been trying to give you some idea of this portion of Utah, but don't know how to do it, unless by representing it as one mass of mountains, with their bases and smaller peaks covered over by the wash of centuries. It is a vast plain with ridges and peaks of mountains breaking through it, and dividing it up into long narrow valleys. The only part of it settled is along the immediate base of the east summit of the Sierra which rises abruptly above the plain without intervening foothills. The next range eastward is the "Gold Kanyon range" in which occurs sulphates of lime and iron, and carbonate of lime (chalk) the latter occurring in hills of considerable size. From this fertil-izing source the lower Carson Valley has been formed, and yet I do not believe a single crop of wheat or grain has ever been put in in this valley. Upper Car-son is formed most from granite debris, and yet produces good crops of wheat. Eagle Valley is made from a basalt in the main Sierra, and wheat does better. But the best portion of the country is yet to be settled.

The old settlers of Carson instead of tending to their farms are principally engaged in trading (i.e. robbing) with the emigrants to California when there are any. When there is no emigration they rob each other; and though they hold each a mile square of beautiful land, a few head of stock and a few acres under cultivation, is all you see that informs you that they are "farmers."

The Mormons (who have lately came in) are doing better. Industrious, quiet and persevering, they have already done more (since last spring) to "start" the country in the right direction than all the rest put together. They are taught to work, and leave speculation to the "elders." But these same "elders" I am afraid, are a precious set of scamps. The Reeses,[30] two of them, know you. I believe one or both were formerly of Utica. They are samples.

Polygamy is the cornerstone of Mormonism—on it is reared the whole superstructure—and of its practical working you can hardly believe too much if you believe the very worst stories in circulation. Even in Carson on the very edge of California we were assured that some Mormons hired out their supernumerary wives to work for other Mormons, and drew their wages and lived off them. We have heard some awful stories.

We are well provided for our trip. We sold our claim for $150—a very fair sale everything considered. Our partner, William Louget, will go with us if he can sell his claim with the house for $100 more. Otherwise he will stay here until February when he will join us, or we him as our trip prospers. He is a member of the other company—and our arrangements with him in regard to what we may find are that he is to be treated as one of the family and put on the same footing in the company which we would do if he was connected with us or not. He is high-minded, honorable, and honest.

We have ample clothing, bedding, and the dearer part of our provisions, together with chemicals, books and apparatus, purchased, and ready to be packed up at Mud Springs or El Dorado as they now call the town.

The Pennsylvania Election news rather surprised us at first, but on carefully going over the ground, we have much cause for congratulations. The rock on which we split there was the treacherous Fillmore leaders, and we trust the Republicans at the convention refused to have anything to do with them. The whole batch of them have been in the market since Fillmore's nomination, buying and selling all who trusted them. Give Pennsylvania to Buchanan rather than join with them on any terms, but we fully believe that if the Republicans have washed their hands of Fillmore they will have carry the state for Frémont.

In this state the Fillmore dodge worked to a charm. I know of but one pro-slavery Know Nothing that voted for Fillmore while the northern fools went him with a rush, and regretted their votes the day after the election. We have seen enough in this canvass to know that if the issue can be fairly brought before the people, the next election will tell a different story than the last. It needed one canvass to get light into the dark corners. The discipline of the Democratic Party seems to make them almost invincible. We must learn from our enemies. Our principles are correct, and come victory or defeat this time success is not far off.

We will put our spare time, when we get settled in Carson, on our motor, of which, the more we examine it the more sanguine we are. But let it succeed or fail, we are prepared for everything.

Our post office will remain as heretofore, Mud Springs. Give our warmest love to all.

Truly and affectionately your sons,
E. A. and H. B. Grosh

◡◠ Based on handwriting it appears this letter is from Allen Grosh:

December 2, 1856
Sugar Loaf, California
Dear Father,

By date of this you will see that we have not gone over the mountains. A sudden storm the night before we were going to start threw snow on the mountains as low down as Placerville, rendering the mountains impassable except with snow shoes. Indeed the winter is unprecedentedly early, and sets in with a vigor which promises abundance of water. We will try to go again in February.

We have modified our motor into a rotary engine very easy . . . you know the result . . . by next mail. We have means sufficient for this experiment, and will have, by the time we are ready for it, sufficient to test it with the pressure engine. We thank you for your proffered aid, but shall not need it. Your kind expressions of sympathy were also right welcome. But we think the worst is past. Our outfit for Carson, gives us an abundance for this winter—and we will pass it comfortably.

We would like you to subscribe for us for the *Weekly Tribune.*[31] We have no means of sending the money, except through the mail and gold coin does not pass current through the hands of Democratic postmasters in California.

Truly and affectionately your sons,
E. A. and H. B. Grosh

1857 AND 1858

Letter from Allen Grosh:

January 13, 1857
Sugar Loaf, California
Dear Father,

The place in which we live is so shut out from the rest of the world, and there is so much sameness and monotony in our lives, that we do not keep so close a watch on time as we should do. The last two mails for the Atlantic closed before we were aware of them. This must account for our silence.

The rains this winter are late and very light. We have been ever since water came busy "opening" the claim on which we are at work. It has proved a much more tedious and laborious job than we expected. However, it is now done—finished today—and from this time we hope to do a little something. Should we have more rain this month—which we expect—we will probably do very well. But quien saba?[1]

We have made but little headway on our motor—though not from want of trying. After any number of attempts to cast lead pipe we are at present brought to a stand still for want of money—a chronic complaint!—to purchase the same. A week or two will remove this obstacle. We expended our money in laying in a stock of provisions for the winter, and as we made a considerable saving thereby we have no cause to complain. But could we have known the amount of time and labor necessary for the opening of our claim we would certainly have made provisions for it.

We are divided in opinion as to the probability of our getting over the mountains this winter. Hosea thinks we will, but I am very doubtful. The month of February is generally warm and the weather settled. But I fear this year it will be stormy. The cold spells are from three or four days to a couple of weeks duration, during this and last have been followed by so little rain that I fear it is only putting it off until February.

We have considerable that we must take over with us—chemicals, etc.—otherwise we could go at any time. I would mention that there is with ordinary caution little or no risk to life in crossing the Sierra Nevada in winter. We may lose everything we have with us, but that will be the extent of the damage. The snow lies extraordinarily deep on the mountains this winter.

January 24th

We had no opportunity for sending this letter to town—and the water coming, had not time to do it ourselves, in time for last mail. A neighbor is going over tomorrow and for fear we will be too busy to go ourselves we send it by him.

Hosea has probably made quite a "hit" in chemistry. He has improved the Mexican mode of working silver so as to prevent a loss of quicksilver—and in fact can dispense with quicksilver altogether. He only completed his first experiment this afternoon. It worked beautiful all through. He will try it again, on a little larger scale in a day or two. I have not time to describe the process now, but it promises to be of great value to us, saving from 75¢ to $2.25 on every pound of silver, besides saving in time probably more than one half. We will keep it secret until we firmly establish ourselves. The Mexican process requires no fuel.

(Since writing the above Hosea has commenced copying from Regnault's "Elements of Chemistry"[2] the Mexican method of working silver and will explain his method in connection.)

I see in the last *Tribune* a change of the judge at Genoa (Mormon Station), Utah Territory. We had not heard of it before. One error needs correction—the statute against lewdness quoted is not a U.S. but a Utah territorial law. While in Carson Valley we examined their "revised statutes" as carefully as our time and business would allow. Their whole aim and purpose is to throw power into the hands of the "elders" and dust in the eyes of the "gentiles." The Governor and courts have extraordinary discretionary powers, and if the general government had but half done its duty, much evil would have been prevented. Some of their laws are clearly in violation of the constitution of the United States and Congress should look to it. They have what they call a "Perpetual Emigration Fund." Their published laws give no act creating or incorporating such a fund. It is clearly a <u>church</u> fund. Yet certain fines and forfeitures are directed by law to be paid to said fund. We noticed also that in an act ordering the collation of the laws of the territory they order the publishing of an accompanying list of laws "except such as are abrogated, obsolete, not in force or <u>not necessary to print</u>" and in the list they mark sections and parts of acts and laws "not print" and refer by page and volume—not to the printed laws but to the statutes probably kept at the secretary's office at Salt Lake City, so that an outside "gentile" over curious must go there for information. But <u>why</u> not necessary to print?

Their laws on slavery are for outside show—very fair, just and reasonable, but dependent altogether on the <u>magistrate</u> for execution. They allow boys of 18 to vote which law works to the advantage of the Mormons in mixed settlements. They grant exclusive right to elders, etc., to water courses timberlands and what not, making the "small fry" of the church dependent on them for water for irrigation, water power and wood. Property when <u>first</u> brought into the territory is taxed extra, thereby placing the "extra" on all property passing <u>through</u> Mormondom, in the possession of "gentile" emigrants to California, etc. They strengthen the power of courts, magistrates and elders by discouraging the settlement of good lawyers in the territory by taking away all right to collect a fee. The consequence the territory is filled with "bush lawyers" a disgrace to their profession and the territory. Common law is done away with by forbidding the citing of precedents from any other courts than their own. The laws of Utah, if extended over Carson Valley, will invalidate the right of every squatter to three-fourths of his claim, and compel them to go to Salt Lake even for their cattle brand, besides confusing and unsettling all their customs and usages. They have no law of marriage, common law not being recognized by their statutes. Justices of the Peace and other town officers are elected by the people, but none of their acts are legal until the governor "commissions" them.

These are a few of the faults we found with their laws, now hastily recalled to memory. I would mention before leaving them that the laws but in one case recognizes the plurality of wives and in that case it would not excite suspicion at first reading. But polygamy rests almost entirely on the right of "illegitimate children <u>and their mothers</u>" to inherit the same as the wife and legitimate children—"the court" being the judge of the child's parentage. We could fill pages with complaints against their <u>published</u> laws. But of those "<u>not print</u>" statutes we can only hear of through the emigrants skinned and fleeced at Salt Lake. From their stories the "not print" list is pretty long and heavy.

The object of the elders is clear. Their aim is to build up an aristocracy as strong and hideous as the oligarchy of our Southern states. By appropriating to themselves "exclusive control" of water privileges and valuable woodland, "regulating" the "use" of the first and the "roads" through the last, receiving fees and collecting tolls thereon they have made great strides towards making permanent their power. Their "wives" are <u>nothing more than slaves</u>. They <u>support</u> their husbands in most cases. Their people are ignorant, fanatical and devoted to their elders. They are taught to look on all "gentiles" as enemies and the deep undisguised contempt manifested toward them by all Americans brought in contact with them confirms this belief. But I have one incident to relate which I think reliable, and must give it room. A girl, accompanied a family (relatives) into the territory—an <u>old</u> "elder" wants her for a supernumerary "wife"—she rejected his offer with horror and disgust. For this she was turned from the house of her relatives. From door to door she sought employment, and everyone was closed against her. Starvation stared her in the face. She was compelled to "marry" him.

Truly your son,
E. A. Grosh

The following appears to be written by Hosea Grosh:

The Mexican process . . . is as follows, . . .

The ores are first reduced to a fine powder and made into heaps called "tourtes," containing 500 to 600 hundredweight, on platforms of stone. After being moistened with water containing from 2 to 5 percent of common salt, are stamped by horses or mules. In a few days about ½ or 1 percent of "magistral," consisting of a roasted copper pyrite, containing 8 to 10 percent of sulphate of copper, (blue vitriol). It is again stamped and the first portion of mercury added. When this has been well disseminated through the mass, a small portion of the material is washed in a wooden bowl, to judge by its appearance, of the progress of the operation. If the surface of the amalgam is grayish and the mercury gathers easily the operation is progressing properly. But if it is greatly divided, and its surface exhibits dark spots, the magistral is in excess, and lime must be added to decompose it or there would be a great loss of mercury. If it retains its fluidity the chemical reactions do not advance, and more magistral must be added. After the first portion of mercury has combined with silver enough to form a doughy amalgam a second portion is added and afterwards a third portion. The whole process lasts 2 or 3 months according to the ore and temperature.

The chemical reactions are as follows.

The suphate of copper, of the magistral, and the salt (chloride of sodium) mutually decompose each other forming protochloride of copper, and sulphate of copper decomposes the protochloride of copper, and by causing it to become subchloride, is itself converted to chloride of silver. The sub chloride of copper reacts on the sulphret of silver, forming sulphuret of copper, and chloride of silver, the chlorides all dissolving in the solution of salt. The mercury in its turn reacts on the chloride of silver, changing it to metallic silver, and is itself changed to subchloride of mercury, (calomel) while the metallic silver combines with the rest of the mercury. It is important that no free proto chloride of copper remains, for it would increase the loss of mercury, by parting with its surplus chlorine to it and converting it to subchloride of mercury, (calomel). The loss of mercury is from 1 to 3 parts to every part of silver obtained.

We intend to dispense with the mercury, lixiviate[3] with a saturated solution of common salt and precipitate by iron, which will throw down the copper and silver which are easily separated. We need not fear, therefore, any bad effects from an excess of magistral, and can considerably hasten the process, by adding more than can be used when there is mercury present.

✐ *Based on handwriting it appears this letter is from Allen Grosh:*

March 2, 1857
Sugar Loaf, California
Dear Father,

We received your long letter of January 9th, by the last mail and thank you for the trouble and pains you have taken in regard to the "Utah Enterprise." It is more liberal to us than we asked but it is probably best so. We approve of the names throughout. Mother's family we of course wished to be regarded as our own.

We still receive favorable opinions in regard to the ore and the situation of the mine. An old and very dear friend—George Brown[4] an old bachelor like ourselves—acts as our agent, so that everything is safe. The winter has been very severe in the mountains but we think we will be able to pass over early in April.

Your letter gave us good news of all the family except Uncle Moses. How sorry we feel for him. A cheerless boyhood, a hard apprenticeship—blows instead of kind words—force in place of persuasion—kicked and knocked about by drunken brutes—and that hot temper of his checked only sufficient to distort and deform its growth, what wonder he looked not straight at life and selected a false foundation to build upon? "Dog and cat" typified his idea of the world. Has it not been fearfully realized in his life? While we sorrow for him we thank God that you and all the rest thrive and are happy.

Our mining this winter has proven an utter failure so far—how it is almost impossible to say. We have very bad luck. However we do not despair of doing well yet, before we leave. As it is we'll have to leave our "Motor" until after we get our silver mine opened and tried. Failing the silver we will go below to San Francisco.

Times are hard on the miners generally. California is nearly at the end of her rope. Our love to all.

Affectionately your sons,
E. A. and H. B. Grosh

Letter from Allen Grosh:

April 2, 1857
Sugar Loaf, California
Dear Sister,

We received your very welcome letter of February 8th, by the last mail.

You know by this time that we did not get across the mountains last February—and right sorry we are that we could not for we have only thrown away our time by staying here. For the first time since we have been in California our spring rains have failed us. The consequence is that we have thrown away our whole winter's work. Last fall, when the snows came down on the mountains, and stopped our passing over, we had laid in what provisions we counted on taking over, and some $75 in cash. This money we also laid out in provisions. We have sunk all this, and now come out "just even" that is, not in debt! We got our claim—by a great deal of work—in good working order for "ground sluicing." Had half the usual quantity of rain fallen during the past month, we would have made from $200 to $500 clear, which would have given us all that we could desire—sufficient to determine fully the value of our silver mines. However we think we can now make a little with the rocker, and with that little . . . Go we must, and go we will, and that as soon as the snows will let us. But it is worrying and annoying, this continued bad luck of ours. Fortunately we have got all our chemicals, and apparatus, and probably after the first month we can make the mine pay all expenses while we continue our prospecting—that is, if it is near what it promises, and one month's labor will determine this last point. By the time you receive this, we think that we may be at work on it, and it won't take many weeks to determine if all our troubles are over. We could very easy raise as much money as we want, and ten times more, probably, by disposing of part of our interest, for several old silver miners have pronounced so favorably on the rock that many on their recommendation have solicited us to sell an interest or admit them into the company, they paying for the admission handsomely. We have declined all offers of this kind, coming as they have from persons objectionable or strangers.

Should our silver fail we will try our hand at something else—not gold mining, you may rest assured of that! We think of going, in that case to San Francisco, but it is all undetermined. We may stay in Utah. The climate of Utah agrees with us admirably. A year or two spent there would make us healthy men.

California is now beginning to put on her beautiful spring dress. The season is very backward. Usually by this time the hillsides and hollows are colored with flowers—an ocean of flowers, with waves of green, blue, yellow and white. The oaks have all leafed out, and large patches of yellow and orange, sprinkled with blue flowers begin to appear. A week or two more and we will be in the midst of a vast flower garden.

We live on a high and commanding ridge which commands a view of the Coast Mountains almost their entire length, as well as the snowy summits of the Sierra and a part of the San Bernardino Mountains. Nearly the whole of the Sacramento and San Joaquin valleys are taken in a single glance of the eye! The air is so transparent that a mountain a hundred miles away is distinctly visible.

While winter visited you with a cold stern countenance, he passed us with a smile. Parts of December and January were what we here call "cold"—the thermometer say at 24° for the lowest. The thermometer stood at 74° on the 1st of February and has been up to that figure quite frequently since. But the want of rains has kept back vegetation and the miners' hopes, together. The miner does not enjoy beautiful weather. The patter of the rain on his canvas roof is to him a song of joy, and makes him happier than would a deluge of sunlight or miles of flowers. Though I like the music of the rain drops right well, I must say that the glories of a California spring are to me inexpressibly dear. The soft air, the bright sunlight, is to me a perfect realization of spring.

We wait with anxiety that bond of agreement—and hope that it may have come by this mail. If not, the next will be in time enough. We wish to have it with us when we go over. Our partner during the past year—William Louget— will probably go with us. Like us, he has heretofore been one of the right "unlucky ones." Mr. Brown will meet us there.

Give our warmest love to all—and kiss our nieces and nephews for us. Lord how old we are getting.

Truly and affectionately,
Your brother,
E. A. Grosh

✑ Letter from Allen Grosh:

May 2, 1857
Sugar Loaf, California
Dear Father,

Yours of March 13th we received today together with the "articles of Agreement" etc. We will hardly start across the mountains before the middle of the month, and may be not before the 1st of June. Though we have had a very dry winter of it here, yet large quantities of snow fell on the summit, and we will not go until we can cross with animals, when we will be able to take over our provisions with us. We are cramped for means, and a month's living there now, at the high price we would have to pay for our provisions, until we could get them from California, would destroy us. We are very sanguine that within 6 weeks from the time we get there we will be making $50 per week, with the rude machinery we will at first make use of, and have plenty of time for prospecting, and fully examining the veins.

What you write of Uncle Franklin's health excites uneasiness, and we will wait with anxiety for your next letter. Consumption is almost unknown in this part of California and I think that all tendency toward it in either of us is entirely gone. We are almost constantly bilious and subject to frequent light attacks of a low fever, more disagreeable than dangerous. Cayenne generally will remove them.

We were pleased to hear of Warren's determination to abandon school teaching as a profession—a thankless profession it is in this age. Hosea has a horror of school teaching, and I think that if he takes to pick and shovel he will find it healthier, more profitable, and full as pleasant.

Judge Taney's decisions has been received by the few Republicans with whom we have exchanged opinions, with feelings similar to your own.[5] The "opinion" is, in itself, so absurd, and contradictory that, had it been delivered in Congress as a "speech" it would have damned its author even with his own party. We can draw this consolation from it: it is straightforward and bold, and looks only one way, and we now have something substantial to fight. The "monster" has left the bush, and we think the people of the North must look at it. They can shun it no longer. The struggle is to the death, Slavery or Freedom must die. And the struggle, bold and open, must waken even the slow blood of Pennsylvania.

California politics are "all a muddle." What is to become of the state, God only knows! It seems as though we had turned all our blessings into curses. All good seems to turn sour here. Give our love to all and believe us to be affectionately,

Your sons,
E. A. and H. B. Grosh

~ *Letter from Allen Grosh:*

May 20, 1857
Sugar Loaf, El Dorado County, California,
We do hereby authorize and empower E. A. and A. B. Grosh to strike our names from the Articles of agreement of the Utah Enterprise.
John M Butler
William Louget

(Second Sheet)

We think we know to what grandfather most objects in the Articles. The power placed in our hands to buy and sell. We will not abuse this power. While we hunt others will do so also. This power may enable us to prevent much trouble to the Enterprise by litigation from conflicting claims, and this is the reason for our asking this power. While things of this kind are unsettled, in a new country much can be done to prevent future trouble. As soon as we secure everything, it is our intention to surrender the Articles of Agreement and build up a good, strong working company out of the Enterprise, and place everything in their hands.

The two names above were added to the Enterprise. They are members of the Frank Company.[6] They misunderstood the terms on which we were willing to admit them. And with their consent we struck off their names. Give love to all.

Truly and affectionately your son,

E. A. Grosh

Letter from Allen Grosh:

May 21, 1857
Sugar Loaf, California
Dear Father,

After many and vexatious delays, we at last get off. We start tomorrow, accompanied by William Louget, and a friend—a Mormon—who has just come over from Carson. I say a Mormon—though I think he is rather sick of them.

To raise means sufficient for the trip we have been compelled to impose on your good nature, and have given an order on you for $50 as security. This $50 we intended as a gift and should never have thought otherwise of it but for your offer last fall, and we have thus pledged it only after much hesitation, and with the greatest reluctance. We do not think that you will be called on for it, but you may if we fail in Carson, or in case the holder, John Hise[7] of Pittsburg, Pennsylvania, returns home before our return, which is barely possible. Should we fail with the silver mine and not be able to make up the amount we will let you know a mail before the order is sent you, and make up as much of the sum as we possibly can.

The past winter has been one of bitter disappointment and vexation. In looking back to it we can discover nothing to blame ourselves for and therefore do not regret what we could not help—but we do not want to go through another like it. We will hold on to the claim. It is well opened . . . , but we do not think we will touch it again. Should the silver fail will probably go to San Francisco.

Your letter of April 12th filled us with uneasing and sorrow. It is true then that Uncle Frank must be cut down in the strength of his usefulness? The world at its present crisis needs good health and willing hands, and here is one of the best and straightest fading from the sight of his friends, and they no power to save. Uncles Frank and Pinckney stand to us in the light of elder brothers rather than uncles, and as such we have long regarded them. We still have some slight hope that your next letter will give us more cheering news.

We sympathize deeply with you and mother in all your troubles. They are indeed trying. I cannot understand how it is that a clergyman cannot speak on a moral principle for which he has struggled all his life, simply, because, in [the] march of time, that principle has become a political one. If the Catholic and Mormon priesthood drag their sects bodily into the political arena and launch the thunders of the church at all amongst them who refuse to worship and sustain slavery, what right have the "Democracy" to complain when liberal Christians raise their voices, as Christians, against the Iniquity which [is] now threatening to flood our country with whips and chains, and blot out the sun of Liberty with black clouds of slavery. For the past dozen years the Catholic and Mormon . . . have been working hand in hand with "Democrats" openly and avowedly; and I believe there was not a paper belonging to either of these churches but what advocated—and that, too, earnestly—the election of James Buchanan during the last campaign. "Birds of a feather," etc. But you are right, father in saying that the brunt of the fight should fall to the hands of the young. Veterans like you have commenced the fight, we are to finish it, and God grant that we do our work as well as you have done it. Indeed, when I look into the future I gather new courage, for the signs of success are everywhere cheering. What though men deny the principles on which our government rests? They cannot alter them. What though the great idea of Protestantism is condemned?—Religious Liberty cannot be crushed out. What though men turn from the orthodox church and God to their counting house and ledger? They cannot forget the Christ who died for them—they must look to the God he revealed to them. The light of his age is too great—they must open their eyes and see Him. A new age—a new world—is surely dawning. Common schools and the newspapers begin to do their work. The shadows [of] ignorance begin to fade before the light of reason and truth, and error and wrong can find no safety in the darkness. Can they stand in the Light?

. . . to disabuse the mind of grandfather concerning the Enterprise. The laws and rules for holding claims on mineral veins are vague and uncertain. The whole matter is generally left to those in a section who are directly interested in the matter. A "claim" is from 150 to 1000 feet to the person. We adopted, for the sake of avoiding all trouble, the rule of the Sacramento Quartz Convention—150 feet. The "claim" under this rule is too small for individual action, and association has to be resorted to. In forming the Enterprise we have only chosen our associates—and we have naturally chosen friends and relatives instead of strangers. Again—by a series of accidents, (which if we succeed, we will regard as almost miraculous,) we think we have discovered the "key"—so to speak—of the system of silver veins in the neighborhood of Gold Canon. The whole success of the Enterprise must rely on our knowledge and industry—we have therefore taken good care to secure liberal compensation in case of success. It is not a speculative movement but an industrial one. Silver mining, is a sure occupation resembling in that feature coal mining much more than gold mining. It is a fit business to build upon. Our object is to settle all our family eventually in West Utah. Those who will not come, can sell out to those who will. If the mines are as we expect we can do all this. Some few that cannot be induced to emigrate may make money out of it without doing anything. Better they have it than strangers. But the real value of the mine must be produced by careful and painstaking labor. Not on the mine, alone but on the country. He who commands both the currency and the exports of a country can either destroy that country or build it up. Hosea and I will have that power if we succeed. We feel the responsibility—and do not think we will abuse it. Our habits are simple—our desires are simple—in money matters we are thoughtful and we really believe prudent. In this matter since we have left you we have walked through fire. We think our habits fixed, here we feel positive. Should God place in our hands great wealth, we will strive to do our duty as . . .

↶ *Based on handwriting it appears this letter is from Allen Grosh:*

June 8, 1857
Gold Canon, Utah Territory
Dear Father,

We left Mud Springs on the morning of May 23rd and arrived here on the 27th, at midday. We were parts of two days in snow, but did not have much trouble from it. We struck the vein without difficulty, but find some in tracing it. We have followed two chutes down the hill, have a third traced positively, and feel pretty sure that there is a fourth. The two chutes we have traced give strong evidence of being surface veins. We find no connection between the "black rock" or iron, and a vein beneath. The ends, as we have exposed them exhibit the shape of a v, thus: [*see the figure on the left*] A perfect leaf of quartz passes all around the vein or bed. The following is a diagram of the set of veins: [*see the figure on the right*] A. seems to be the centre from which all seem to radiate. B. we have traced by boulders. C. we have struck the end. D. the same. E. is suppositions, though the evidence of its existence is tolerably strong. B. A. C. may be the true vein, and the chutes D. A. C. may be superficial spurs. Still, we shall hold them all to be surface veins until we have proof to the contrary. We have pounded up some of each variety of rock and set it to work by the Mexican process,[8] and it may take two weeks before we get a result. Meanwhile we have gone to gold washing, and have every prospect of making more than our expenses. We commenced this morning and made over $2 out of 60 buckets of dirt. We can wash 200 per day. We could determine the vein much quicker, by other processes, but any other than the Mexican would require a furnace and crucibles, which would consume at least a week's time, at dead expense.[9] By waiting we can use that time profitably, and we wish to make everything . . . as we can. Besides, in case of failure of the silver, we can have enough to repay what we borrowed, without doubt, so you can make your mind easy on that point. The rock of the vein looks beautiful, is very soft and will work remarkably easy. The show of metallic silver produced by exploding it in damp gunpowder is very promising. This is the only test that we have yet applied. The rock is iron, and its colors are violet blue, indigo blue; blue black and greenish black. It differs very much from that in the Frank vein—the vein we discovered last fall. The Frank vein will require considerable capital to start. The rock is very hard and the vein very much split up. Our vein lays very compact, so far as we have examined it—not a leaf of foreign rock in it.

If the vein turns out half as well as we have reason to expect we will at once put up a wind mill. Should it fail wholly or in part we think we will stay this side the mountain until fall, as we can do very fair at gold digging. We have secured good claims. We took up a couple of claims last fall—and they were very fair ones. On our return we found that the miners had respected them. Brown, our agent for holding the veins, took charge of them for us. But it is very rare to have a mining claim respected under like circumstances.

Gold Canon is at present mad with the prospecting fever. Two or three parties will soon start out in search of rich gold diggings, reported to have been struck during the prospecting rage in California in '50/51 and 52. There is no telling what will be the result.

We sent with our last the wrong copy of the Articles. Please let Mr. Smith give us power to sign for him.

Mr. Louget returns to California tomorrow and we send this hasty note by him. We will try and let you have full information, etc., by 1st of July steamer. In haste your sons A. and H.

❧ *Letter from Allen Grosh:*

August 16, 1857
Gold Canon. Utah Territory
Dear Father,

We have delayed writing for the simple reason that we did not know what to write. We were completely lost in the mist. Things looked so very bright that [we] were suspicious, and it took much time and labor to clear away the fog. Right glad are we that we were prudent, otherwise we must have awakened hopes only to be disappointed.

The rock composing our vein is one which presents more difficulties to the miner and metallurgist than any other are containing silver that we have heard of. It is almost impossible to reduce it by heat to a liquid slag, and its want of a sufficient quantity of sulphrin makes roasting useless. It is not as we had supposed, the magnetic oxide of iron, but the magnetic sulphuret of iron—a single sulphuret to which the sulphin sticks with great tenacity while it "eats up"—to use the expressive phrase of an old silver miner—almost all the flux you can add. It is a rock of great importance from the fact of its being the outcrop of many silver mines in Utah, California and New Mexico and until our metallurgists become fully acquainted with it, [it] will be the source of ruin and trouble to many. A humid assay of the result left in the crucible by smelting the rock with litharge[10]—the surest mode of testing refractory silver ores—is, we believe, in the hands of an ordinary chemist an impossibility, whereas antimony occurs in the rock in any considerable quantities as is the case with ours. By turning to any chemistry you will see that chloride of antimony does not form in the same manner as the chloride of silver, lead, etc. But it is only formed by passing a current of chlorine over the antimony. You will also see that the chloride of antimony bears little or no resemblance to the chloride of silver. Yet the result from the smelting of this rock with litharge, dissolved in nitric acid will throw down a precipitate so closely resembling chloride of silver as to be readily taken for it. Nor is this all. This precipitate redissolved in boiling hydrochloric acid will deposit on the cooling of the acid in the form of small crystals, just as does the chloride of silver. From this solution will precipitate and do compose it by iron, mercury, and the amalgam bears a . . . close resemblance to the amalgam of silver, except that you cannot retort it on a shovel from a tendency it has to jump and fly about—a phenomenon we have frequently observed in a much less degree in gold amalgam, when the heat is applied too suddenly. Further; though this deposit does not darken in the light, yet there generally being some silver in the rock, which converted to chloride, darkens readily [and] gives to a careless observer the phenomenon of being turned dark by light.

Our first assay was half an ounce of rock, melted with litharge, carbonate soda and pulverized charcoal.[11] The result was $3,500.00 per ton! by hurried assay, which was altogether too much of a good thing. To guard against getting by accident an assay high above average we resolved on testing a considerable quantity of rock. After many failures we at last succeeded in getting the silver in part, we thought out of about 20 pounds of rock. Though our after experience inclines us to believe the method we used would not even have started the silver: the lead we used became hard, and had all the appearance of an alloy of lead and silver and by hurried assay contained nearly as much silver as our first assay. After four attempts we succeeded in cupelling it, imperfectly, and got <u>nothing</u>, though our awkwardness and blunders would have lost considerable silver had it been present. However, we got the hang of cupelling pretty thoroughly which will be of great use to us hereafter. We now hit on the mode for bringing our rock to a perfectly liquid slag. We took half an ounce of black rock (our vein), half an ounce white quartz, two and a half ounces carbonate soda and one ounce litharge, and melted in a crucible. The making of this crucible we would mention, caused us weeks of experimenting. The last excess of soda and the litharge would combine with almost anything we could use. We had used a brasque of charcoal[12] and clay for the protection of the inner surface of an iron kettle in our former smeltings. This brasque, the charcoal pretty well burnt out, we pounded up and mixed with clay and wood ashes in the proportions of 2 parts brusque, 1 part clay and 1 part wood ashes. The crucible answered perfectly and the smelting was equally satisfactory. This we cupelled with a hand bellows, by piling charcoal over the cupel. Unfortunately the antimony (or whatever else it may be) in the rock caused the lead to scatter and divide into little globules through the slag and we tried to gather them together with borax, in which we only partially succeeded; and on placing the lead in the cupel we carried a small portion of borax with it. Here the borax combined with the litharge and formed a slag so that we lost the true result. However, we satisfied ourselves that the result obtained by nitric acid was entirely false.

We will not attempt to give you any idea of the toil and trouble this error caused us. Day after day and week after week we were at it from daylight to dark, hanging over glowing furnaces, and the thermometer in the "nineties." However, we have mastered the rock and now know something about it. It has caused us this inconvenience that we had to drop everything else until we mastered the error. The alloy of lead and antimony so closely resembles that of lead and silver that to our certain knowledge at least half a dozen experienced silver miners were deceived by it and in this regard this same kind of rock has given our California metallurgists considerable trouble. A sample brought from Lower California they pronounced worth some $800 per ton, but they "could do nothing with it." One Mexican silver miner only, of all that saw the rock, gave us a true account of the rock. His opinion was that it was the outcrop of a very rich mine of silver, but contained only a small quantity of silver itself.

We have determined, almost positively, that the crossing of two silver veins, has caused the large quantity of iron ore to gather as it has at this point. The Frank vein we had struck many hundred yards from this crossing and it exhibits rich specimens of galena. Another vein (almost certainly a spur of our vein or the Frank) we have struck just across the creek, does the same. We assayed a small quantity of the galena from the last named vein by cupellation—the result was 5/1000 silver—about $200 per ton—besides the lead. We have several other veins, which are as yet untouched. We have great hopes and are very sanguine of success.

Sunday, August 23rd

We have written the forgoing hastily, and at spells. On the determination of the black rock, we immediately set about getting ready for gold digging. This required considerable preparation, as we had to prospect and get ready to haul dirt, from where we might find it to the nearest water. Wednesday Hosea hurt his foot seriously by striking it on the inner side just under the instep with a pick. He is now doing very well, and has suffered but little pain, considering the severity of the wound. It confines him to his bed. He will probably be helpless for three or four weeks to come. He bears it all with his usual patience, and talks about how much worse it might have been had it struck a little further back or a little further front!

In addition to this misfortune we have had another. George Brown, our most intimate friend this side the mountains, and one of the few right good men in Utah, has been murdered at Gravelly Ford on the Humboldt, by a train of blood thirsty, Indian-murdering Arkansans and charged by them with cattle stealing.[13] He, and two others, Thomas Jones and John Jones, were there as traders with the immigrants. So many accounts of the affair are in circulation that it is as yet impossible to get at the truth. The most probable story is this. Buster's train the murderers commenced killing Indians indiscriminately near Fort Laramie, and followed up the business until they came to Brown's Post— (some say that their atrocities got the U.S. troops at Laramie after them, and that Buster only escaped by going on ahead of his train.) At Brown's post they turned out their cattle, without watch or guard.

Gravelly Ford is right in the nest of the Shoshone country, and for the past few years the Shoshones have been quite troublesome—for which, by the bye, the Mormons are generally held responsible by most of the settlers here. During the night the Indians killed several head of stock—six or seven—Buster and his party accused Brown and his party of being in league with the Indians, and fell on them and murdered them. Buster's train it would appear were not altogether at the massacre . . . report says, that shortly after a part of the train came to the post, when one of them recognized poor Brown, and told the murderers that they had killed one of the best men in Utah, and that he surely could not be guilty. The train hastened on rapidly to California by the Truckee route, fearing to pass through Carson. Report says that they killed another man—a trader—after murdering Brown and his party. One thing is sure, they could never have got through Carson alive. Tom Jones was the son of one of our settlers. John Jones was an apostate Mormon. He was compelled to leave Salt Lake last spring, as Brigham's destroying angels[14] were after him. Tom Jones bears a good character—the other was a stranger here. But Brown's death has excited universal indignation and the charge against his character is as universally disbelieved. In fact no man could bear a higher character than poor Brown. In a community no way distinguished for virtue, he stood like Paul in the congregation of Israel, and all bore testimony to his worth and integrity. What property he had was with him, so near as we can ascertain, and the post was sacked and robbed by Buster and his ruffians. Brown was a bachelor and a native of Rhode Island—Providence we think—and lived some time in New York. He was a small man, 5 feet 4 or 5 in height, light hair, light blue eyes—from 40 to 45 years of age and a shoemaker by trade, fond of travelling and a great wanderer. He came to California in 1849 returned to the states in '50, and came back again in '51. The first trip was through Mexico and the last by way of Panama. He came to Utah (Carson Valley) in the spring of 1853. We send these particulars, and request that you will write to some of your acquaintances in Providence, Rhode Island, and make inquiry concerning his relations. He is a member of the Frank Company and was to have joined us in the enterprise, this fall, furnishing us the necessary means, whenever we pronounced the enterprise veins sure . . . He offered us last spring any means that we might want and to spend the summer with us in prospecting. But we thought that we could do the work alone just as well, and in case of failure it would cost him his usual summer's work trading on the road, which with gold mining in the decline was his occupation. We therefore told him so, and as his means were limited to a few hundred dollars, which he would need in buying stock from the emigrants, we concluded to get along as best we could until his return when his stock would be as available as money for what little we might want. We once sent for him to come in, being deceived by our false prospect, and on discovery of our error, terminated the message. He most probably never received either letter.

Brown's death, and the accident to Hosea's foot will make it necessary for us to call on you for that $50. We will send word with this to Mr. Hise to collect it of you. We wish you to write at once if you are panicked by our call on you. For we can repay you before the closing of the mountains and will do so if it causes you any trouble. If you can spare it without inconvenience we will not attempt to repay you until we are sure we have enough to lay in a full winter's supply of provisions.

Our present arrangements are these. A young man by the name of Galphin[15] has a quartz vein of gold. He, Hosea, and I have made this agreement. Of the three, one works on the quartz vein, and the other two at gold digging until the value of the quartz vein is determined. Should the vein prove rich enough, we will put up a windmill to work an arrastra. Should it fail or should we have doubts of its paying, we continue our gold digging, so long as we chose to work together. Should the vein pay we have the use of the vein and mill with him for one year, all three of us sharing, equally. He offers us two-thirds of the vein, but we have declined, as an interest would bind us to give it our full attention, whereas now we leave it when we choose. Our only object is to make a raise sufficient to lay in a full supply of provisions for the winter and furnish what little else may be necessary to prospect our vein as we are pretty well satisfied that they will not fail us. Should we not be able to lay in our winter's provisions we will winter in California.

We have our cart and road nearly ready—a few hours work more and everything will be ready. We cart downhill about 400 yards, and have dirt that will pay from 4 to 8 cents per bucket. Two of us can each dig and wash from 200 to 300 buckets per day after we get a going! Hosea not being able to work, I will cart and wash alone and will probably do about 100 buckets per day. We will not touch our silver rock (except to put some 40 or 50 lbs of the black rock already ground through by our improved Mexican process, for an accurate test) until we can clearly see our way through next winter.

We will in relation to poor Brown's death, gather what information we can, and then, as soon as we have a little means that we can spare we will put his case in the hands of one of the best detective officers in California, a personal acquaintance. We are bound to go to the bottom.

Love to all.

Affectionately your sons,
E. A. and H. B. Grosh

᥯ *Letter from Allen Grosh:*

September 7, 1857[16]
Gold Canon
Dear Father,

I take up my pen with a heavy heart, for I have sad news to send you. God has seen fit in his perfect wisdom and goodness to call Hosea, the patient, the good, the gentle to join his mother in another and a better world than this. In the first burst of my sorrow, I complained bitterly of the dispensation which deprived me of what I held most dear of all the world, and I thought it most hard that he should be called away, just as we had fair hopes of realizing what we had labored for so hard for so many years. But when I reflected how well an upright life had prepared him for the next, and what a debt of gratitude I owed to God in blessing me for so many years with so dear a companion, I became calm, and bowed my head in resignation. Oh Father, <u>Thy</u> will, and not mine, be done. Our happy faith in the perfection of God's wisdom and goodness will be your consolation as this cloud passes over your head, for well I know your heart is full of the great hope which caused Paul to shout in triumph, "O death, where is thy sting! Oh grave, where is thy victory!"[17]

At the time of his death I had gone to see a physician in Eagle Valley,[18] some 14 or 15 miles from here. It was very sudden—unexpected but very peaceful. Not a shudder, not a gasp, not a change of feature marked the parting of soul and body. He simply fell asleep. It was such a death as God blesses the good with.

The immediate cause of his death was the wound in the foot I mentioned in my last [by first letter] it occurred about the middle of the forenoon of Wednesday, August 19th, or Thursday the 20th. He died Wednesday, September 2nd. We were packing dirt from a small ravine to the right fork of the main canon. I dug and Hosea drove the jack. We had brought no water with us, for drinking and becoming thirsty (it was very hot) I started down to the main ravine for a drink. I met Hosea as he was coming back for another load and told him what I was going for and that I had not quite a load dug for him. On my return he was setting on the ground beside the dirt holding his left foot in his hand. "I have done it now," he shouted as I came within hearing, and on my asking what he had done, he said that he had "struck the pick into his foot" "Why how in the world did you do it?" I asked as I first saw the wound. It was a frightful gash. The dirt we were digging was only 16 or 18 inches deep, and, though it dug hard, there were but few stones in it. He smiled, and said that he hardly knew how he did it. He then pointed to a large quartz rock laying loose on top of the ground just on the edge of the hole. "Somehow" he said "I hit that." He would not let me carry him to the house but rode the jack. The ground was rough, and the jolting caused him considerable pain.

For about a week it got on finely, in spite of the hot weather. But the evening of the eighth day, his foot was swollen, and the wound was closed. The next morning I lanced the foot in two places and got out considerable matter, which relieved the pain, and checked the swelling. I also changed the poultice from rosin soap to bread and soda. The bread and soda worked very well, and I think that if we had continued it everything would have come out right. In all matters concerning sickness I generally deferred to his opinion. He thought if we only could get Indian-meal to substitute for the bread it would get along the faster. On going down to the store I found a pound or two of old Indian-meal. It was about half bran. That night we tried a poultice of it. The next morning I did not like the look of it and asked him if I had not better go back to the bread again. I thought the bran was too healing. He answered, "Let us try once more and see how that works, I was very feverish all last night, and maybe that is all that is the matter with it." In about two hours, he complained of a strange sensation in the foot. He said that it seemed as if a little ball was underneath the flesh and was running all over his foot. On examining the poultice was dry and the wound was closed. Though we tried the bread and soda again we could not get it to draw right. That night he suffered much pain. From this on I gave up all idea of working thought we had not a cent and were in debt. I might have done so before just as well, for my time was so broken up by loss of sleep, etc. that it amounted to nothing.

Monday afternoon I went down to the store—four or five miles—to see if I could get either opium or laudanum, so that he might get his necessary rest. I could find neither but I saw Mr. Rose,[19] and he told me that he had some at his house in Eagle Valley. He also recommended me to try fresh cow-dung as a poultice. I took some cow-dung up with me, and applied it immediately. I should have mentioned that the leg had commenced swelling, and that we could not check it. The poultice at once checked the swelling, and lessened the pain, and next morning everything was looking well again. I found a man who was going up to Eagle Valley and sent by him for the opium, and also for a little quinine, cayenne and several other things if they could be got. I could get nothing here and Hosea was quite bilious besides touched with the dyspepsia[20]—the result of his confinement to bed. I understood the man would be back that evening; but that evening found that I was mistaken. This evening also occurred the mishap which I think sent Hosea out of this world. The cat jumped on the bed, and in doing so lit with all his weight on poor Hosea's sore foot. It caused him intense pain. That night he suffered great pain, and next morning he had a high nervous fever, accompanied with . . . complained that during the night he had been slightly flighty. He was very cool and calm, and before I went to see Dr. King (formerly of Deerfield, New York and with whom we had some slight acquaintance) we had considerable conversation. He said, that "through God's mercy we had passed through as great trials as this—and to that mercy we must trust—without God's mercy what would we be? Dear Brother! He spoke as though the trial was as much on me as on him. He was so uncomplaining and made so little of his sufferings that it took close watching to see how sick he really was. After some little thought he consented to a proposition I made to send to you for $50 or $100 so that we could, on the strength of it secure the services of Dr. Daggett,[21] the only good doctor in Carson Valley, should they be necessary. Little did either of us dream of the danger being so near at hand.

I dressed his foot. It was rather cold. He quieted my apprehensions by saying that it was the effect of the warm poultices. The poultice was warmer, a little, then blood heat. He felt it very sensible, and we both congratulated ourselves on the favorable symptom, as the poultice before that had been warmer and he had hardly felt it. He complained of being very sick, just before I left, but felt no other pain.

About 9 a.m. I started for Eagle Valley to see Dr. King and got what medicine I could, leaving him in charge of Mr. Galphins, who came to the house a few minutes after I had left. I had not gone far before a feeling of uneasiness took hold of me. Twice I threw myself down behind a cedar bush, completely overcome with a great dread that it would terminate fatally. I prayed—oh with what agony I prayed that he be spared—that the loss of the limb might be the worst. Finally to get rid of this dreadful apprehension I struck across the mountains, which though it shortened the distance a few miles, was very rough, and I was almost barefooted. Dr. King was very kind to me. He recommended the continuance of the cow dung poultice, as being the best to be had here. He did not regard the swelling of the foot and leg—neither the coldness—as anything serious. He spoke as if a wound got along very slow in this country, but did not seem to think that the danger was increased thereby. Hosea complained of pain in the back, and one particular spot, near the shoulders on the left side, he said produced nausea if it touched the bed. The doctor regarded it only as the result of the pain and loss of sleep together with slight biliousness. He gave me four pills of Blue Mass[22]—which I took for fear of hurting his feelings. But I got ten or fifteen grains of quinine.

(2nd Sheet)

Though I could get no physic but aloes or Epsom salts, both of which we had and would not use. I regretted very much that I could get no hops, as I had more hope of allaying the nausea with that than anything I could think of. Of Mrs. Rose I got some opium and a few ounces of garden peppers. I started back with a lightened heart. It was just dark as I got back. Mr. Galphin met me a few steps from the house. "You must prepare yourself for bad news, Allen," he said. I heard strange voices in the cabin, and I thought that either Dr. Daggert, or some physician travelling across the plains had come on to the Canon and had been sent up by the miners below, (as Hosea was thought a great deal of) and that it might have been pronounced necessary to amputate the foot. I was quite unprepared for the answer to my "what it is?" "Hosea is dead!" Oh the terrible force of that blow! Oh! The utter desolation of that hour. What I said and what I thought I will not say. The world was nothing to me and I envied the dead. But Mr. Galphin pointed out the necessity of courage and resignation and as he told me of his quiet, peaceful death, and I began to gather consolation from the blessed faith Christ died to give to prove to us. Galphin is a Campbellite,[23] and a South Carolinian, and was very prejudiced against Universalists and Republicans. Hosea had conquer his prejudice in a very great degree, against both, and he was much struck with his peaceful happy death. "As I sat by him" he said in telling me of it "as he breathed his last I could not entertain a doubt—he surely has gone to Heaven."

When I left I forgot to set some peppermint tea I had made for him, within his reach. He was getting up for it as Mr. Galphin came in, who told him to lie still, and handed it to him. He drank pretty heartily of it, and in a few minutes threw it up. Mr. Galphin then heated some water at his request, and gave him a couple of spoonfuls, which remained on the stomach. About 2 hours after he had occasion to get up and Mr. Galphin helped him, he complained much of nausea and weakness. He had sunk very much since I had left. He said as he lay down "this country doesn't suit me" and Mr. Galphin thinks he added "and I am going to leave it." Mr. Galphin thought that it was best for him not to talk, He therefore walked into the other room, and laid down on the bed, so that he could see him. Hosea remained perfectly quiet for some time, and Mr. Galphin thinks that he was asleep. An hour or more he heard Hosea breathe hard and went to him, and spoke to him. Hosea heard him, for he partially turned his head to him, and opened his eyes, but did not fix them on him. His eyes closed again as if he had fallen asleep. His breath shortened easily and without effort—"it died away" to use Mr. Galphin's expression. His features wore an expression of happy pleasant, sleep, and with his last breath he did not even stretch himself. "He fell asleep." Let us thank our heavenly Father, for even as He envelopes us in a cloud of sorrow His mercy shines through it.

Hosea loved this country very much, and that morning, even, had talked of the enjoyment of the coming fall. We both concluded that it was best for him not to think of working or gold digging anymore, but spend his time hunting on the jack, and also examine several points for copper and silver, things in which he took great delight. Mr. Galphin is firmly persuaded that he alluded to leaving this world in his last words and has regretted over and over again that he did not pursue the conversation. Had he had any idea of Hosea's critical situation, he would have done so. But he had failed very much since I had left, and had slept but little the night before, and Mr. Galphin thought if left quiet, he would fall asleep.

It gives me pleasure to state to you that the possibility of one or both of leaving this world had several times been the subject of full and free conversation between us. We had lived so much together, with and for each other that it was our earnest desire that we might pass out of the world as we had passed through it—hand in hand. His hope in the faith you had taught us had robbed death of all his terrors to his mind. Earnest and truthful, patient and hopeful, he was to me a guide and a support such as few men had ever enjoyed. One mind and one purpose actuated us in all we did.

Friday Evening, the 11th.

What I had written has been done as several times. I had written it very calmly, and the pauses have been occasioned by my being busy at work—to which I rushed as the best relief to my feelings.

The miners buried Hosea very respectably and gave me many marks of their kindness and sympathy. I send a lock of his hair. He wore it as you do yours.

I should say that the cause of our being without means was the murder of George Brown. He had tendered us money when he left in June. But we declined for this reason—the money would be of use to him in buying stock . . . , and in the fall his credit would furnish . . . the black rock presented so many difficulties that we lost patience,[24] and, relying on Brown, we dropped everything determined to master it. The very day we had determined it we heard the first rumor of his murder. We could have made all our expenses at gold digging, by giving a good portion of our time to that object. We had good claims—very good—but they are at this season a quarter of a mile from water. We were packing ore from them when the accident happened to brother. More for the purpose of prospecting than anything else, calculating to cart the dirt should it prove rich enough. Since his death I have been prospecting and working with Mr. Galphin. Next week I shall go back again, having the cart built and everything ready. I am, of course, behind hand, probably $50 or $60. I can square it all off, if I do not have any back set, by the first of next month. If I can get ahead sufficient to lay in my . . . groceries and clothing—I will winter here. If not I will put all the time I can onto the silver veins, and try and determine their value in part, and then pass to California, in which case I shall probably winter in San Francisco. I feel very lonely, and miss Hosea very much—so much so that at times I am strongly tempted to abandon everything and leave the country forever, cowardly as such a course would be. But I shall go on—it is my duty and I cannot bear to give anything up until I bring it to a conclusion. By Hosea's death you fall heir to his share in the enterprise. We have, so far, four veins—three of them promise much.

This has been a very bloody season, six persons belonging to the Canon have been murdered since we came over, besides one person shot on the Canon who did not belong to it. The first case was a desperate negro who stole a horse and was followed. He was armed with a gun loaded with a handful of pistol balls. He was very daring and desperate and it was resolved to make short work with him, but before he was shot down he killed an Irishman belonging to the Canon. This occurred at the mouth of the Canon.[25] The next was Brown and his two companions all of the Canon. This week John McMarlin (formerly of Fort Plain, New York) and James Williams were shot from their horses crossing the Sierra Nevada. They both belonged [to] the canon.[26] Six persons killed—besides the negro—out of a class of our population—miner's—numbering less than fifty! The Irishman was a hard working industrious man. Brown was one of the best men in Utah, one of the Jones' was a blacksmith, and bore an excellent character—the other Jones was a short time only on the Canon and left Salt Lake with the "destroying angels" . . . Williams was our constable, and cannot be replaced . . . bold, daring, prudent and an excellent man. McMarlin when he just came on to the Canon—about three years ago—had some things bad about him—but the last year or more had changed for the better and changed greatly—became an altered man. He will be much missed. Besides all this blood shedding the Indians on the Humboldt have been very bad—

Within this last week the Mormons have suddenly commenced selling-out everything they have got, and are all going back to Salt Lake.[27] I have much to tell you, for Carson Valley is full of news, but my thoughts are with you at the time that you shall receive this sad letter. You will have no care to hear and I have no care to write, except of dear Hosea. Yet when we think of it, he has only gone to join his mother. Of the six, five yet remain to you—one has gone to her. May God deal gently with you and yours, dear father, and temper your grief with the bright hope of our glorious faith. Surely it is a rock on which to build, and as I took my last look of what remained of him I thanked God with all my heart that he had, through you, led me to such a building place. As I thank you, father, for the dear faith you have taught me, so will all your children as they learn that one of their band has broken the fetters of mortality and flown <u>Home</u>! With a thankful heart I send love to all.

Truly and affectionately, your son,

E. A. Grosh

❧ *Letter from A. B. Grosh (the father of Allen and Hosea):*

Returned from California
Andersonburg, Perry County, Pennsylvania.
October 25, 1857.
My dear—<u>dear</u> Allen—

Your two letters, of August 16-23, and September 7-11, both came to hand yesterday—forwarded from Perry, New York, and were opened and read in due order. I have no words that will describe our feelings of grief and sorrow at the news contained in the latter. I read only the first lines, and feeling utterly unable to control my feelings or voice, uttered the words, "Hosea is dead!" and leaving the letter for your uncle, retired to our room, whither mother soon followed me, and we wept long and sadly together, before we found sufficient composure to return to the sitting room and complete its perusal, with many pauses for tears. We thank you much and fervently for the minuteness of detail, for the spontaneous and artless manner of utterance and description, and above all, for the frequent references to the sources whence you drew support and consolation, and to which you have so affectionately pointed us. Yes, God is good—infinitely and universally good—and though His ways and thoughts are so high above ours, yet even where much is hidden from our finite and imperfect understandings—veiled in clouds of afflictions and sorrow—yet even then and there it is frequently manifested as the silver edge that, shows the lining of the cloud! All my children have not only been spared until the youngest has reached manhood, but they have been spared in general health and physical strength, and spared from the ravages of vice, to become virtuous, intelligent and useful men and women. Notwithstanding my infirmities of temper in early life, and my ignorance and neglect in their full and proper education, they have outgrown my errors and defects in their training and have fulfilled the better desires of my riper and better age. And you, my dear Allen, who inherited much disease from the medical maltreatment of your mother,[28] to heighten your own nervousness, and on whom, as the oldest born, fell the heaviest portion of my neglect and errors in your training—and whose impulsiveness made your mother and I unwilling that you should go to that ill-fated land of gold, until your brother decided to accompany you—I cannot but thank God that Hosea was spared so long—for seven long years of exile, sickness, toil, hardship and suffering—until, I trust, you have assimilated by <u>self-education</u> to a great share of his strong caution, prudence, and patient, cheerful, steady philosophy, combined with your own greater energy and zeal.

Oh, how often have we mourned over your united failures, disappointments and misfortunes; and hoped, almost against hope, that the tide might yet turn, and bring you both back again, to our arms. How often has my soft bed felt hard, as I thought of the hard one you probably pressed—or my warm clothing, and abundant fare seemed sinful, as I reflected on your wants and sufferings. The idea that at the very time poor Hosea lay on his dying bed, and you toiled and watched alternately, without a cent, and without the necessaries for the well, let alone comforts for the patient, cheerful dying one—that at that very time we were reveling in abundance of everything, spending money freely in travel and luxuries—how this idea <u>will</u> creep in and intrude itself, adding poignancy to our grief and bitterness to our sorrows! Would that we could have shared them with you—or that you had been here to share them with us. And even now, as I think of your necessities, to which I have means to administer, I know not how to have those means securely reach you.

That $50 due Mr. Hise will be gladly met when his request reaches me with directions how to get it to him, as soon as the present financial derangements will enable me to procure available means. I will write to Perry, New York, by the next mail, and have my agent procure and place in the bank there, that amount, in readiness for his draft, should it be presented at that bank—or to be forwarded to him, as he may direct, at my order. Should he send here, it must be by letter to me, and I have that amount in father's hands, or can borrow it from your uncle, and send it by return of mail.

And now, for yourself. I hope you may be able to stay in Carson's Valley; for I fear that when the crash of our failures[29] reaches San Francisco, it will be but a poor place in which to find employment at wages during the winter. If you can contrive any way to turn an order on me, payable next spring, into a reasonable amount of cash, you may draw on me for $50, or even $100 more, and I will meet it in a month after sight or notice. (I most heartily wish you had done this as soon as you and Hosea needed it!) And I have thought it probably that you may get such an order cashed, through some Universalist who knows me by reputation—as, Brother Edmunds, editor of the *Star of the Pacific* at Nevada City; or Hon. Mr. Washburn,[30] the delegate to the Philadelphia National Republican Convention that nominated Frémont—who is one of the Washburn brothers of Maine, a Universalist family. I mention these, should all others fail.

There is something very painful in the idea of your remaining in Utah or California, independent of distance and dangers, arising from your utter loneliness there. I wish you could but realize one or two thousand dollars. That would be sufficient to make a start in business here which would yield you a good living, and more, with half the toil and deprivations and none of the solitariness and dangers you endure there. In the present state of markets I expect to purchase some 20 or 25 acres of good land in Lancaster or Chester county, next summer, for less than $2,000, and put Warren on it to garden and farm it in the summer and teach school in the winter of the first year—then with the proceeds start keeping fowls, bees, etc., and (having got things ready) go on with the produce raising in earnest. He is tired of teaching because it furnishes no immediate prospect of means toward marriage, and is confident that with 10 acres he can make money much faster. He shall have a fair trial on double the quantity; for mother may need a home there someday, and means to make a living also. He is now teaching in Irishtown (above Marietta)[31] for 6 months at $38 per month and intends saving $150 out of it, which, with present funds, will give him $200 to start with. He will begin when everything is at the lowest prices—for so they will be a year hence—and $200 cash will go as far as

$300 now, or $400 a year ago. If he teaches another winter before beginning (as he may do) he can easily have $350. Fruits of every kind, and garden produce of all kinds, sell well and readily in every large village, and Lancaster and Philadelphia furnish steady markets. His labors will be less hard than mining, and more sure, with no privations (physical, social or religious) worth naming. He is to pay me the interest, and I pay the improvements he makes, until we become joint-owners and joint-workers, or either purchases the whole. By next summer I expect to be able to pay ¾ths of the purchase money, and the rest may remain on the land. As to wealth, it never was so desirable to me as the comfort and society of my children, and now that its pursuit has cost me the life of my noble Hosea, its value has decreased immeasurably in my eyes. I would cheerfully relinquish all his and my share in the enterprise ever can be worth, to have you comfortable and contented at my side for life. Such are my wishes and feelings; but I am well aware that you may see matters and feel responsibilities differently, and I, therefore, bid you "God speed" most heartily in your pursuit of what your judgement and conscience dictate.

A letter from Letitia yesterday, says they all have had severe influenza, but are now perfectly well. The weather is fine, but business dull, almost dead, at Oswego;[32] though Cyrus does not feel the pressure much, as his employer can easily stand it. A letter from Malvina, at Utica, a few days ago, says that her health is still slowly improving—her recovery from each backset carrying her a little further healthward, and the attacks of prostration being further apart. I have feared, very much, that she would be the first called to rejoin her mother; but hope she may be spared to us for some years yet. The death of Hosea will be a severe trial to her health as well as her feelings. I have copied all in your last letter relating to the event, for Emma and Letitia. Emma and her family, and our friends generally, were also well when she wrote. A letter from Pinkney, received yesterday, says all are in usual health there. Charley is to visit Blaine next Thursday.[33] From other friends I have not heard for some time. Here all are in usual good general health except the Dr.[34] He is slowly and very gradually losing flesh and strength under his disease, until he has ceased, pretty much, to leave the house yard. His very sore throat & hoarseness prevents conversation except in a hoarse whisper and with some pain; but he seems perfectly aware of, and prepared for the result—and that result may occur before next spring. Our future is yet uncertain—our stay here an experiment. While the Dr. lives we remain; and probably a year longer to settle his affairs, and give a fair trial to living on the farm. If we then conclude to remain, it will be until Blaine can take its management—if we go, we take the 2 younger children as our own (receiving pay for their actual expenses) and will try to settle down in some little home we can call our own. So we intend, if God wills. Meanwhile I am working daily—sometimes pretty hard—

on farm and garden, and about the house. Have planted over 400 strawberry vines, and as many peach and plum stones, grubbing up the hard clay and mellowing the places thoroughly. Thus I am taking my lessons and serving my apprenticeship to my new business. How much pleasanter would be my labors, if I could expect you to join me in a year or two, to settle down with us, and make and embellish a home for us all. Oh my heart is sore with deferred longings to embrace my absent sons—and sorer still, now that one can never return to these mortal arms!

Your requests about Mr. Brown shall be attended to shortly. I have also made a note for Malvina in relation to Mr. McMarlen of Fort Plain, as she will call there on her way. Franklin replied to Pinkney for me yesterday, but I will write soon to Warren, and send him your letter for perusal and safe return. I believe I have every letter received from you and Hosea.

I wrote you in May, of our contemplated removal here this fall—from your addressing me at Perry, New York, I fear you have not received it. I also wrote you in June (I believe), and in July, and since my arrival here—which was September 2nd, the day of Hosea's death—every month.

The lock of Hosea's hair was not in the letter—I hope it is not lost—if not, do not fail to send it yet. I wish we had a daguerreotype of him taken lately; and we want yours also, as soon as you can send it. All send much love and blessings.

God bless you. Affectionately your father,
A. B. Grosh

∾ *Letter from A. B. Grosh:*

Andersonburg, Perry County, Pennsylvania.
November 20, 1857
My dear Allen,

Without waiting for another letter from you I write again—the second since hearing of the death of our dear Hosea—and I grieve that I must now, in my turn be the communicator of very sad news to you—of news that, in part at least, will be as unexpected to you, if not more so, than yours in regard to Hosea's departure was to us—as unexpected to you as it was stunning and afflictive to us.

Pinkney's son Charlie came up on a visit to Blaine, (as I wrote in my last he intended doing,) but after he had been here about a week only—on the 5th inst. Warren came as a special messenger to summon him home. We could not credit the news—supposed a mistake in the name—when it was announced that "Aunt Lizzie"[35] was no more on earth. Had it been grandfather,[36] as we at first supposed, or even Uncle Pinkney, whose health had been poor all fall, we had not been so surprised—but we had left Lizzie, less than a month previous, in good health—heartier than she had been for years—and only a week before, Charlie had left her as well. We left next morning, (Warren, Charlie and I,) and reached Marietta that night at 10, and remained to the funeral on Sunday, 8th inst.

It appears that she had taken a cold a few days previous by sitting in a draft of cool air, but visited on Monday evening until 5, and attended to indoor household duties on Tuesday until 8 p.m., when she retired complaining of stitch in the side and difficulty of breathing and coughing. On Wednesday morning Dr. Stehman (partner of Dr. Armor, their Homeopath physicians, now of Lancaster) was called in, and in the evening Dr. Armor came and prescribed. (It appears that Dr. Armor apprehended that she could not survive over 36 hours, but contrary to his usual practice, he did not inform the family, designing to call again next day.) Pinkney seated at her bedside gave the medicines promptly, reading to and conversing with her until 10 p.m., when she seemed, and said she was much better—could breathe easy, etc. She then fell asleep, and Pinkney drew closer to the bed, and observed from time to time how very evenly and quietly she slept, until about 11, when her medicine due. Feeling loath to disturb such an easy sleep, he delayed giving it. She breathed noiselessly, so that he did not hear her, but observed the rising and falling of the bed covering, until about 11:15 when he went and got the medicine ready, and approached her with it, and—she had ceased breathing! The smile was on her lips, and the form relaxed as before—the hands and feet were warm, but she was gone! Emma and Nettie were immediately called, and Warren coming in from Libhart's at that moment, ran to the stable and took the horse barebacked and procured Dr. Stehman immediately, (who happened to remain in town) but she was gone beyond the reach of medical art and kind affections. Father was not aware of what had occurred until next morning on the stairs he met Warren, who was afraid he might fall in one of his fits, and therefore merely told him Aunt Lizzie was much worse. Poor Pinkney, stunned and crazed with the sudden blow, had not said a word or shed tear after the Dr. left, until father met him downstairs, and said, "Warren tells me Lizzie is much worse," when he merely said, "she is dead," and burst into tears. Father's tears and wailings were piteous to hear; for such had been her constant attention and quiet kindness to his wants, that he had become very strongly attached to her—but they aided much in unburdening Pinkney's overcharged feelings.

They clothed her as in life, and she remained unchanged in appearance (though decomposition probably commenced soon) until Sunday morning—the same smile on her lips, and looking as if in easy sleep. The immediate cause of her easy death was probably the same as in Hosea's case—a translation of the inflammation from the lungs and pleura to the <u>heart</u>. I send you an *Ambr.*[37] containing portions of your last letter, and a brief biographic sketch of Hosea, and will send you another containing an account of our dear Lizzie's death and burial. Enclosed you will find one from the *Lancaster Daily Evening Express*.

And now I proceed to another sad event, but not so very unexpected to us, and for which my previous letters have already prepared you. When I left here for Marietta the Dr. was able to be about the house and yard, and seldom even laid down to rest during the day. It was agreed that I should remain until Tuesday, if needed—especially as Alexander and family were coming up for a week's visit the next day. And as father and family were lonesome after the friends had all left on Sunday, I concluded to remain. But I felt uneasy that the change to sultry weather would affect the Dr. very unpleasantly, and that I might be needed as <u>physician</u> if not as nurse—and on Monday spoke of it, and talked of returning in the evening train so as to gain half a day. But I had a severe sore throat, and, towards evening, diarrhea set in, and so I gave it up, till next morning. At 10½ p.m. I was roused out of bed by a messenger who came with the news that our dear Franklin had left us at 5:45 a.m. (Monday), and was to be buried at 10 a.m. on Wednesday. Warren got a team, and at midnight we left through rain and wind for Elizabethtown, where we arrived at 2½, in time for the express train, and reached here about 11 a.m. on Tuesday. It appears that the change of weather affected him on Saturday evening so that he went to bed early—but ate and slept pretty much as usual on Sunday, though he did not get up: On Sunday evening he told the hired girl[38] quietly to prepare <u>two extra</u> candles for lighting, as they would be needed. At about 5 a.m., Monday, he rung the bell, and the family were called. His throat had been so diseased that for some weeks he spoke only in a whisper, and at times was so sore, that even whispering was painfully difficult. He now spoke not, but by looks and signs showed consciousness—and shortly after, stretched out his frame and folded his arms across his breast and closed his eyes, and thus gradually, easily and calmly breathed his life away—dying without a struggle or a groan, or any signs of pain. Three deaths in our family circle in a little more than two months—all easy and painless—ages, respectively, Hosea 31 years, 10 days—Lizzie 34 years, lacking 14 days—and Franklin 39 years, 10 months, and 5 days. The *Amb'r.* I will send you containing Lizzie's death, will also contain a notice of the Doctor's.

On Wednesday after services in the house we carried him to Blain, 4 miles west, and deposited his remains alongside those of Mary and their infant son, and daughter Belle.

By his will Alexander and I are left joint executors and guardians of his children (the two youngest to be constantly ours), with a request that arrangements be made for my living on and having charge of the farm also, if Alexander and I can agree on the terms. But, independently of the possibility that Alexander and I may not agree, the death of Lizzie, and father's bereavement of Franklin, may change all our plans and calculations. It is not the first time that we have been taught that "it is not in man that walketh to direct his steps."[39]

Father has written us a long letter, stating his strong desire to have his only two remaining sons and their families near him during the remainder of his life, and proposing that as soon as I can be spared from settling the Dr's. affairs, we shall come there with the children, and all make one family in the old home. To enable us to do this, he proposes to give up the lumber business entirely to Pinkney and I (capital about 9,000 dollars); he reserving only the rents of his real estate, house and furniture for his own support, and to pay us for his boarding and washing.

This plan of father's requires not only much consideration as a business matter, but also consultation with Pinkney and family, and many domestic considerations. Its main object—the union of the families, or, at least, bringing them near to each other, is very desirable to us all, if it can be effected without too much sacrifice of interest and comfort, and duty to others.

We intend remaining here until spring, at least—probably longer stay may be necessary—further we make no positive calculations, but follow the indications of a wiser and better will than our own.

We have not heard from Oswego since my last to you. From Utica, Malvina wrote about 2 weeks ago that they were all in good health but herself—she had rather retrograded, but was then improving again. I wrote her to leave for Philadelphia as early as she could, to escape the coming cold weather, and get better medical aid than she was then under. Since then we have not heard from or of her; but hope she will write as soon as she reaches Philadelphia. My health is very good, and mother's also, except her chronic complaint and a large boil on her arm.

Pinkney's health was pretty good—better than usual—when I left there; and the family were all well, except Emma Montgomery who had a bad cold. Father's health was as usual—his fits of falling unconsciously, are without pain of any kind, caused probably by ossification or paralysis of the valves of the heart. He continues talking while falling, and when he recovers, which is in a second or two, he continues the conversation where he broke off. The news that the Mormons intended returning to Carson, for a general massacre of the Gentiles, has given us some uneasiness about your remaining there. Be careful and wary of your person and your life. I send you a Rhode Island (Providence) paper, with your notice of the murder of Mr. Brown in it. All unite in much love to you. Heaven bless, preserve and prosper you, and return you speedily to us again.

Very Affectionately, your father,

A. B. Grosh

∾ *Letter from Allen Grosh:*

> *Envelope postmarked:* Michigan Bluff, California. December 17
> *Envelope addressed:* F. J. Hoover, Esq. El Dorado, California.
> *Written on envelope:* Please send a line to me Michigan Bluff, as it may be a long while before I get out. If I get them, your names may be useful in procuring passage in the stage to friends and medical treatment. I received only one letter in Carson Valley—from January. The others miscarried.
> Preserve this letter carefully. A. B. G.

Blind Ravine, Middle American River
December 12, 1857[40]
Dear Governor and Friend:

We were snowed in in crossing the Sierra Nevada, and escaped only with our lives. We were forced to take to the water courses as our guides to our way on account of the thickness of the snow storm. Our matches got wet, and both barrels of the shotgun clogged, and we were obliged to seek shelter from the wind and storm on lower ground, where we succeeded in saving our lives by burying ourselves in the snow. The next day, the storm still continuing as violent as ever we were compelled to follow the river as the only means of knowing our direction. For four nights we were without fire and too cold to weave snow shoes. We buried our feet and legs in snow and will probably save them. We are now in the kindest of hands and are treated as though we were relatives rather than strangers. We have been taken into the house of an Italian and a Chilean, and the only trouble we find is to prevent them doing too much for us. We are indebted to them so much already, and in fact to the whole settlement, that we feel that we can never repay them. God bless them all until we can make a better return!

My companion is a young Englishman, and though not yet twenty-one years of age has shown the high Norman blood, in encountering our difficulties and trials which has enabled that race to keep its position as the governers of England even into the present day.[41] His name is Morris Buck of Sinai, Canada West.[42]

Morris's feet are worse than mine, as he thawed them out by the fire, while I wrapped mine up in blankets. I had not the energy to rub in the snow, for besides being without fire for four nights we were without food for three days.

The next day we washed in beef brine, and also the next, when a kind Mexican came to our relief with a bottle of mustard, and a liniment of the composition of which I know nothing except that I can detect the smell of oil of turpentine and camphor. After washing our feet in warm water and mustard, he bound up our feet in a white lye poultice. This morning we washed off the poultice and anointed with sweet oil, and our feet look very well. I think that in three or four days I will start for the Sugar Loaf. If Morris is able to go along, he will go with me—if not I shall leave him behind at the best place that I can find, and push on in hopes of getting something to do that I can work at until my feet get well so as to sustain us both, if it is necessary. My Mexican doctor gives me great hope of a speedy recovery to the full use of my feet, but I fear he is altogether too sanguine. My plan is to begin my way towards Sugar Loaf and stop wherever I can get my board for my work. If I cannot get work, I may have to push on as far as Mud Springs, where I <u>must</u> rely on you or some of the Sugar Loaf boys either for something to do or the means of subsistence. Alpha Oak a tradesman at Diamond Springs once sent word that he would pay Hosea and I $25, which he considered our due—If you should go there tell him I would receive any part we would give me very thankfully.

The ends of my fingers are very slightly touched with frost and my wits end with fever, so that I feel very much ashamed at the confused note I have written you. But it must go as I cannot write another. Neither of us can walk. Our friends will try and have us packed out to Michigan Bluff, through the snow. We are 6 or 7 miles above Last Chance and 18 or 20 above the Bluff. We have had a <u>very narrow</u> escape of it. God's providence alone saved us. With this confused account I must put you off until we meet, until then, goodbye all,

E. A. Grosh

Billy Louget had full power to dispose of our share of the house. If he or Al Burns done so it is all right but if they didn't and Billy does not rush over there please let John hold on to them for me and much oblige.

E. A. Grosh
Preserve carefully—A. B. G.
Allen died Dec. 19, 1857 [*written in pencil*]
F. J. Hoover, Esq.
El Dorado
El Dorado County
California

∽ *Letter from W. J. Harrison*

December 20, 1857
Last Chance, Placer County, California
Mr. J. H. Hoover
Sir:

It becomes my painful duty to inform you of the death of your friend E. A. Grosh. He died on yesterday at 20 minutes past 4 o'clock a.m. I believe Mr. Grosh wrote you from the place where he was first taken in from the snow, the day following the date of your letter we the miners by great exertions brought them over the mountain to this place. Mr. Grosh failed gradually from the time we got him in to his death, he exhausted all physical ability and nothing could have saved him. He was frozen up to his knees and so soon as his feet thawed out, the unhealthy flow of blood was too strong for physical strength. You can feel assured Sir that I done all in conjunction with others that could be done for the unfortunate men. Mr. Grosh's companion is yet alive. He is a strong man and may get over it to some extent.

I herewith send you a letter to Mr. Grosh's father; I wish you if you know his address to forward it as I am ignorant of his place of residence.

If you desire to communicate to me direct to Michigan Bluff, Placer County. Mr. Grosh was buried very gently today at ½ past four o'clock.

I remain respectfully yours,
W. J. Harrison

∽ *Letter from W. J. Harrison to A. B. Grosh; this letter is apparently no longer extant, but the partial text, transcribed by A. B. Grosh, appeared in the obituary for Allen Grosh in the "Christian Ambassador," February 20, 1858:*

December 20, 1857
Last Chance, Placer County, California

About the 20th of November he, in company with another young man, left (Bigler Lake, on) the east side of the Nevada mountains, for the more congenial climate on the west side. After they had traveled some 30 or 40 miles their pack animal was stolen by some Indians and white men, supposed to be Mormons. They pursued the trail, and on the fifth day succeeded in recovering their animal.[43] The snow was failing very fast, and they hastened to cross the first summit of the mountain. By the time they were across, the snow was so deep that their jack could not travel. They halted for a few days in a small valley, in the hope of a change of weather; but there was no modification. They then killed their animal, and started with as much of the meat as they were able to pack on their backs. As they ascended the mountain, the snow got very deep, and impeded their progress very much. They at last crossed the second summit; but they had nothing to eat, their matches were wet, and they found it impossible to get a fire during the night. They traveled four days without food, and slept in the snow four nights without fire. On the fifth day they found a camp of Spaniards (Mexicans) who live about ten miles higher up in the mountain. The Spaniards took them in and treated them as well as they could.[44]

. . .

So soon as they came in we procured every comfort that could be had for their benefit. From the time your son arrived here to his death he sunk gradually. Nothing kept him alive from the time of his crossing the last summit and his getting in, but his great resolution and mental energy. His constitution appeared to be impaired by his life in this country, but from nothing but its hardships and privations . . .[45] He had exhausted all physical ability, and nothing could have saved him. He was frozen up to his knees, and so soon as his legs thawed out, the unhealthy flow of blood was too great for his physical strength. His companion is yet alive—he is a strong man, and may yet recover to some extent.

. . .

Could money had purchased the life of your son, it would have been freely paid. . . . The miners wish me to express to you and your family, their warmest sympathy in your present bereavement. They feel sympathy, for your son died in their midst, where all around looks desolate and gloomy, among these inhospitable mountains on the frontier. And in conclusion, sir, please accept the warmest sympathy of a brother "of the mystic tie."[46] I remain,

Your sincerely,
W. J. Harrison[47]

Letter from Francis J. Hoover to A. B. Grosh:

December 31, 1857
El Dorado, El Dorado County, California
Dear Sir,

Accustomed as you have been through life's journey, to rely upon, and seek comfort and support in moments of trial or great affliction from that Almighty arm which is "Strong to deliver," and always willing to succor those who call in the hour of their great need; yet there are trials in life which are of such a crushing nature that unless supported by that power which "never breaks the bruised reed," would render life here insupportable, and a state of misery, rather than a blessing or a state of enjoyment. But much as you have been comforted or sustained in former afflictions; it is now necessary to ask a "double portion of that Spirit" which descends with "Elijah's Mantle,"[48] to sustain you in this last of your great afflictions.

But a few short months ago you received the sad news of the untimely death of a highly talented and beloved son; But O, how painfully true it is, that, "trouble never comes alone."

As your friend, and the friend and neighbor of your truly lamented sons; it becomes my melancholy and painful duty that of transmitting to you the heartrending intelligence of the death of E. Allen Grosh. The enclosed letter one of which was sent to me, and I suppose was the last ever penned by poor Allen, and another from W. J. Harrison giving an account of his death and burial, will give you all the information in my possession of the sad event.

Allen's letter to me was written on the 12th of December and received by me on the 19th or the 20th. I wrote to Michigan Bluff, enclosing money enough to pay his fare to this place, but little did I think that on the same day I sent him aid, he would be consigned to his last resting place on earth. On the 25th I received Mr. Harrison's letter, and on the 26th I wrote to Mr. Harrison requesting him to send me any papers, notes, memoranda or conversation, that might be in his possession, in relation to Allen's operations in Carson Valley; I also gave him your name and address, with a request that he would give you a minute account of the death of poor Allen.

I also return the last letters from you received at this office, the former ones were forwarded to him at Carson Valley but miscarried. I will return to you any letters that may be sent here before you receive this.

I have no definite information in relation to their business on the east side of the mountains; I heard indirectly that, they had discovered two copper and two silver mines; as to their value I have no information.

The following is a copy of a draft given by E. A. and H. B. Grosh, at the time they left here on the last trip to Carson Valley.

Received A. B. Grosh.

Perry, New York

Three days after light please pay to John Hise on order the sum of Fifty Dollars, and charge the same to our account. And much oblige.

Sugar Loaf, El Dorado County, California E. A. & H. B. Grosh

May 16th, 1857

After giving this draft, and funding they were short of means, they give the following due bill. Sugar Loaf, El Dorado County, California, May 20th, 1857.

Due John Hise for money loaned—twenty-five dollars, payable on demand.

E. A. and H. B. Grosh

The draft is still in the hands of Mr. Hise and was held by request of E. A. and H. B. Grosh until they could inform you, or until they would return here, which they can never do.

Mr. Hise wishes you to write him at this office and let him know what you can do in relation to the draft and "due nil."

The money was loaned in good faith and I have no doubt would have been settled had poor Allen have reached here in safety.

The parties here who were interested with E. A. and H. B. Grosh in their operations in Carson Valley have no information on the subject.

I have written to Mr. Harrison about the matter, but have no news from him yet.

I have just received your letter of November 21st, 1857 and herewith return it to you.

And now with the hope that that grace which has heretofore sustained you in the hour of affliction may still comfort and sustain you in this [is] your last great bereavement.

I remain dear sir, very respectfully your friend,
Francis J. Hoover

~ *Letter from C. C. P. Grosh:*

January 29, 1858
Marietta,
Dear Brother Aaron—

I have but sad news to communicate to you at this time; and sad to us all to notify you of it, though you may have received it from some other source. You will see by the extract enclosed, cut out of the *Evening Bulletin*,[49] San Francisco Dec. 29th that poor Allen has gone to meet his brother Hosea, where they will be united in stronger bonds, to part no more forever. How the poor boy must have suffered! If he had only returned after Hosea's death, how we would have rejoiced in each other's company, and been saved from the sorrowful feelings, which his death has caused us all. But God's will be done! I am in but a poor frame of mind to offer consolation to anyone. You must depend upon your own resources and upon that religion which you have taught your children, must you depend for consolation.

The person who sent father the paper is from Lancaster and married to Miss Danner, a niece of father's. She is also from Lancaster. His name F. H. Russel, 114 I Street, Sacramento. He is of the Russel family of Lancaster.

We are all moving along as usual, and all unite their sympathy with you and yours.

Yours affectionate brother,
C. C. P. Grosh

p.s. Father says Mr. Russel will be pleased to give any and every information in his power, if you should wish to address him.

◦~ *Added in the shaky script of Jacob Grosh, the father of Aaron Grosh:*

Mrs. Russel is a amiable niece of my wife, a daughter of George Danner [of] Lancaster.

May God console you all,
Father

◦~ *Letter from Maurice R. Bucke:*

February 10, 1858
Last Chance[50]
Dear Sir,

. . . I wasn't with him all the time and will be days to give you an account of the trip . . . so much to the best of my ability.

On the 15th of November we left Gold Canyon to cross the mountains to California by what is called the Washoe trail, that day got to Washoe Valley 12 miles at two o'clock, camped on the bank of the lake turned out the jack-ass and hunted until night when we came back to the camp the jack[ass] was gone and we did not find him. 16th hunted for the jack all day towards evening found his tracks going to Eagle Valley. I went on to Eagle Valley distance 10 miles and Allen went back to camp, got to Eagle Valley [and] found the jack. 17th got back to camp with the jack one o'clock packed up and crossed Washoe Valley and camped in an old house at the foot of the mountains 8 miles. 18th next morning the jack was gone again hunted about the valley . . . 19th followed the Indians trails about 8 miles . . . about two hours before I got to Eagle Valley a man who was going to Gold Canyon from there took the jack with him supposing that Allen was still living there. I went on to Gold Canyon 12 miles got there about 10 o'clock p.m. found the jack and started back to Washoe Valley got to the first house in Washoe Valley about 4 o'clock a.m. and stopped there the rest of the night. 20th got back to camp about noon . . . packed up and got nearly to the top of the . . . summit 8 miles and camped. 21st crossed . . . Lake Bigler by noon 10 miles traveled around the lake . . . 22nd keep on [north] round the lake until we came to Truckee River which puts out of Lake Bigler. 3 miles went down Truckee River to Squaw Valley. 9 miles. It had been cloudy all day and the night before, and . . . about time before we got to Squaw Valley it began to rain. It rained in the valley where we were but snowed in the mountains that we had to cross . . . we were completely caught, . . . the snow had already fallen to a sufficient depth to completely hide the trail so we had to go back to the valley. 23rd tried it again to cross the summit but the snow falling so deep that we could not keep the trail we found that it would be impossible to get the jack across and so when we got back to camp we killed him and as we were about out of provisions, we lived on the unfortunate jack the rest of the time.

(I had been over the mountains this trail about two months before and on the other side [of] the summit from Squaw Valley had stopped a few days with two men who were herding cattle . . . , when they took their cattle out of the mountains I helped them to drive them down, and what provisions they had not used they left in their cabin which was about 8 miles from Squaw Valley where we now were)

24th snow so hard that we did not attempt to . . . that day I told Allen about the provisions that had been left in the cabin across the summit and that I felt satisfied that if Indians had not taken them that they were there yet. 25th snowed again all day set to work to make snowshoes. 26th snowed again working on snow shoes . . . 28th snowed . . . 29th fine, started to cross the summit climbed all day up the mountain . . . found that we had climbed up the wrong mountain and . . . a deep canon lay between us and the summit, had to go back, about 10 o'clock got back to camp nearly frozen and tired out. 30th fine started early after a very hard day's work succeeded in getting over the summit to the cabin but the Indians had been there and had stolen everything, our matches had all got spoiled in Squaw Valley but we got fire with a gun that we had with us and after roasting some meat and eating it we lay down thankful that we had got that far, from that place the trail ran down a high ridge and . . . it was necessary to have fine weather as the . . . was covered with snow to a considerable depth, we had brought enough meat with us to last about three days thinking that by that time if we found no provisions in the cabin we would get to Robertsons Flat, a place about 20 miles down the ridge. 31st snowed so hard that we could not think of starting out . . . snowed very hard again 32nd . . . cold, started early had about half a pound of meat left got about 5 miles down the ridge began to snow so hard that we could not see fifty yards . . . some time longer until we found ourselves going back in . . . considered what we should do, we . . . almost worn out with the hardships that we had gone through and for the last few days . . . quite unwell, we thought that our best chance would be to leave the ridge and follow down the first stream we came to, that night in trying to light a fire we got the gun stopped up so that we could not use it anymore. It was very cold and we buried ourselves up in the snow to keep from freezing and lay there all night. 3rd snow about two feet deep, snowed very hard followed down a little stream all that day buried up in the snow again at night. 4th about noon got to the middle fork of the American River followed down the river until night, made our bed in the snow again. 5th keep on down the river until we got to where it ran through a narrow rocky canon where we had to leave it climbed up the ridge which at that place is not very high and crossed over to the next stream, when we got there and found no sign of habitation I said to Allen, that we might as well lay there until we died, but he said that as long as he could crawl he would not give up. We crossed the stream . . . Duncan and lay down for the night. 6th We cross the . . . ridge and about noon came by a water ditch where we saw men's tracks in the snow we followed around the ditch a little way and came to some miners houses . . . you can imagine how exhausted we were from the fact that we were half a day walking another crawling ¾ of a mile, . . . we did not know that our feet were frozen but in the days after we got in we could not walk at all.

On the 10th the miners from Last Chance came up and hauled us down on sleighs to this place . . . and then sent for a doctor to Michigan Bluff 14 miles but he did not get here until the 19th. It was then too late. Poor Allen died a little while after he got here, amongst those who were most attentive to us in our sickness was Mr. Harrison and Mr. H. M. Barnes from Trenton Village, New York. On the 1st Dr. Tibbitts came up from Iowa Hill and . . . amputated my feet one about 4 inches above the ankle and the other at the tarsus joint, thanks to the kind attentions of the people of this place I am now doing very well I wrote home last mail for money to take me home, and I should be glad to call on you as I go, if you wish it.

Think of me a friend of your son and a sincere admirer of his character and of his religion.

Maurice R. Bucke

❧ *Excerpt of a letter from Alpheus Bull of the firm Bull, Baker, and Company, San Francisco, to A. B. Grosh; this letter is apparently no longer extant, but the partial text, transcribed by A. B. Grosh, appeared in the obituary in the "Christian Ambassador," February 20, 1858, for Allen Grosh:*

I met your sons on the way to this country in July, 1849, on board the brig *Olga*. Their oneness of purpose was noticed on the ship. When we consider how very closely they were attached to each other while on earth, as mentioned in the *Ambassador*, how very remarkable it is that they were separated so short a time from their union in the spirit-world. Again there is cause for thankfulness to know their end. So many die throughout this country, and not a word do their friends learn concerning them to relieve their hearts of torturing suspense.

APPENDIX A

The Reading California Association

The first letters of the Grosh brothers reflect their preoccupation with the fate of the Reading California Association. Named for their hometown in Pennsylvania, this corporate undertaking unified a group of adventurers as they embarked on an expedition to California's gold country. The following discussion is intended to help the reader keep track of the various people whom the brothers mentioned but also to provide a framework for understanding the sort of cooperatives that sprang up in response to news of gold having been discovered on the West Coast.

On a brisk February day in 1849, thirty-seven men boarded the brig *Newton* in Philadelphia and set sail for Mexico. Eighteen of them were members of the "Reading California Association," often referred to in the letters as the Reading company. All were full of dreams of spending the gold they expected to pick up quickly in the western foothills of the Sierra. It is difficult to imagine that they had thoughts of failure.

As confirmation of the gold in California streams spread through the eastern states, men organized informal companies to head west together. In January, eleven men from Reading, county seat of Berks County, left to join the 150-member Gordon California Association, a company that had previously headed west. While most of the members of that earlier group were from Philadelphia and New York City, there were some from Reading. The townsfolk of Reading turned out to see them off.[1]

Two and a half weeks later, on Sunday, February 12, a public meeting was held at "Dutch John" Ebner's public house on Penn Avenue in the middle of Reading. It was there that the Reading California Association was organized, with Andrew Taylor, aged twenty-three, as president, Benjamin Tyson, treasurer, and Franklin Miller, secretary. Subscribing funds from stockholders, the company held the meeting to elect the members who would go to western gold fields.[2]

California's gold held particular allure for young men, among whom the universal question involved the matter of how to pay for the trip without any immediate financial resources. Many turned to family, neighbors, and friends for backing, with promises to share the buckets of gold waiting to be picked up in California. Participants often drafted formal documents to define each company, or association, and to showcase the percentages investors would receive. Charles Haskins would later compile listings of 35,000 names of "Argonauts," those gold seekers who arrived in 1849 and early 1850. Haskins identifies 350 ship departures from eastern ports or arrivals in San Francisco Bay. George R. Stewart, writing of the California trail, estimated there were 22,500 overland emigrants in 1849.[3] There was no need to record the company papers, so today there can be no comprehensive list of stock companies or associations such as the Reading California Association. The groups were formed as a means to get to California, and there was often little else the venturers had in common. The cooperatives almost always failed, breaking off into smaller groups or as individuals went their own way. Many people never actually mined, finding satisfactory work in other vocations—as carpenters, blacksmiths, masons, merchants, or any number of other trades. California held something for everyone with initiative. Accounting and determining actual profits for that part of a company that remained in the East became impossible in the real world of young California, and most certainly never realized a reasonable—if any—return on the investment.

The Grosh letters document the failure of the Reading California Association, placing the company among those with a similar fate. Once in California, each member went his own way. "The intention was for the members of the association to stick together for at least four years," William Zerbe recalled some years later, "but we disbanded when we arrived in San Francisco. We were apparently not all congenial to each other."[4]

Writing to A. M. Sallade of Reading on March 31, 1850, Thomas Taylor reported: "I have just returned from a visit to the Sandwich islands, whither I was forced to go to recruit my health, which has become impaired by exposure in California during the rainy season. John L. Hahs, Reuben Axe, Henry N. Witman and William II. Zerbe are working at their trade, carpenters, while Uriah Green has engaged in the mercantile business in San Francisco." He said Ethan Allen Grosh was "ill with scurvy" and had been advised to go home.[5] Taylor neglected to mention that he had planned on profiting by bringing back a shipload of vegetables but had been blown badly off course. On his return he learned that the vegetables would have cost more wholesale in the islands than in San Francisco, even before the cost of freight was added.

Two of the Reading California Association died before having a chance to seek gold. Andrew Taylor, like most if not all of the Reading company men, suffered sickness of one kind or another during the travel west. Allen wrote that Taylor, treasurer of the travelers, fell extremely ill at Buena Vista ranch in western Mexico and died in early June 1849. Dr. Walter J. Martin, of Allentown, Lehigh County, became ill in Tampico and suffered varying degrees of sickness from then on. He was still unwell when they reached San Francisco on August 30 and died in that city on October 13 of dysentery. He indicated that he was twenty-one years old when he filed for his passport earlier that year.[6]

When the Reading party arrived, most of the easily recovered placer gold had already been retrieved. For that and various personal reasons, several members returned to Pennsylvania within a few months of arrival. Some did work at mining, but none kept at it as long as the Grosh brothers. Several of the men from Reading were experienced carpenters who readily found work in California, where housing was in short supply.

Charles B. Taylor arrived in San Francisco with the Reading company on August 30, 1849, and soon thereafter booked passage home, leaving September 29. When the census taker stopped by on August 8, 1850, Charles, twenty-two, was unemployed and living at home with his father and three sisters. Family records report that at some point Charles fathered two daughters, who both "died young." He had returned to California by 1880, a widower clerking in a hotel in Stanislaus County, California. In 1900 he had his own restaurant and saloon in that county. Charles Taylor died in Stockton, California, in June 1912.[7]

William Thomas Abbott had also returned to Reading by June 1850. He became a taxidermist at the Smithsonian Institution, securing specimens for the collection. He was a native of England and located in Reading in 1842. His daughter, Hermelinda, born in roughly 1829, married in Reading in 1854. The 1841 census of England indicates he was born in Berkshire, England in 1811, making him the senior member of the Reading California Association.[8]

Thomas Taylor—"Captain Taylor"—the oldest of the three Taylor men, giving an age of twenty-five on his passport, was mining with Uriah Green in Calaveras County shortly after the group arrived. He had been elected captain of the company while they were in San Luis Potosí, Mexico. That he and Allen Grosh were not friends becomes clear in the letters. He turned to farming in the rich soil of the Sierra foothills, and in 1871, he purchased 160 acres of farmland in San Joaquin Township, Stanislaus County.[9] Born in 1824, he died in Stockton, California, on November 20, 1890.

Henry Kerper, who was twenty years old in 1849, went to the El Dorado County mines briefly, then left California on December 1, arriving in Reading on January 16, 1850, with "specimens of California gold." He said ill health caused his early return. Before his California adventure, he had served a three-year apprenticeship as a railroad mechanic. His father had a tannery in downtown Reading. When the elder Kerper died in 1856, Henry took over and expanded the business to 125 tanning vats. In 1858 Henry married Louisa

Reill. In 1872 he applied for a U.S. passport to travel around Europe. In 1882 he sold the tannery, which needed to be moved out of the downtown area. Kerper continued working with a small leather and shoe business in Reading.[10] He was retired in the 1900 census, was a widower in the 1910 census, and does not appear in the 1920 census.

Johnston S. Flack, thirty years old when he left Pennsylvania, returned to Reading in February 1850, "compelled by ill health to abandon the search for gold." He said he was taken sick during the passage out, landed sick in San Francisco, remained sick, and despaired of recovering his health while he remained there. Flack reported that "there was plenty of gold to be had by those able to endure the climate and hard toil of gold miners." He brought along half-eagles made by private coiners in California. The son of Irish immigrant parents, Flack was born in Pennsylvania in 1819. On his return home, he worked as a plasterer in Pennsylvania. He married Ellen, probably in 1852, and they raised several children. Flack committed suicide in 1884.[11]

Samuel H. Klapp was twenty-three in 1849. He quickly tired of wading in cold streams and hunching over a gold pan. He returned home in June 1850 to work as a stone mason in Bern, a small settlement some miles northwest of Reading. He brought with him about $500 in gold dust, which he delivered to the U.S. Mint in Philadelphia. Perhaps in error, the clerk then sent him another depositor's slip, for $1,400. Declining to return it, he was taken into custody in Buffalo, New York, and returned to Philadelphia. There the matter was apparently resolved amicably, since no docket was opened in the federal court in Philadelphia. Born in 1826, he died in 1894.[12]

Simon Seyfert also returned home during the summer of 1850. He was twenty when he joined the Reading company, and following in his father Henry's footsteps, he became an ironmaster in Robeson, west of Reading. He married Ellen, and in time, he and his son Samuel took over Henry's iron manufacturing business in Reading.[13]

Reuben Axe said he was twenty-four in his passport application of 1849. He carpentered in California for awhile, then returned to Pennsylvania during the 1850s and was a master carpenter in Mifflin County in 1860. Reuben was married (Mary) and the father of two boys, born in 1858 and 1859.[14]

Noland H. Witman, age twenty in 1849, remained in Monterey as a carpenter.

John L. Hahs was a widower when he left Reading in 1849, age twenty-eight. He never went prospecting, instead remaining in Sacramento as a successful carpenter. Eventually he bought a farm, married a Spanish woman, and sent for the two children, a girl and a boy, that he had by his first wife. Hahs died in California in roughly 1888.[15]

Peter Rapp, at thirty-three years old in 1849, was one of the older members of the Reading California Association. He left the group on arriving on the West Coast to go to the mines. He was reported on the Yuba River at the end of March. Following in his father's footsteps as a farmer, he cultivated land in Pleasant Valley, Nevada County, near Deer Creek. His younger brother, Henry, came to California to help work the farm. Peter eventually married a woman named Johanna, a native of Württemburg, Germany. In March 1862 she gave birth to twins—Johanna and Barbara. The girls had an older brother, Augustus. Peter Rapp applied to the General Land Office for three parcels of land, which were paid for after his death, by his widow, in 1874 and 1877. Born June 28, 1815, Peter Rapp died on January 16, 1871. He was buried in the Masonic Cemetery in Nevada City.[16]

Robert S. Farrelly was born in February 1824. Instead of attempting to find gold, he first settled in San Jose, then moved to Alameda County, where he took up farming in "Squatterville," which the city of Oakland annexed during his lifetime. He remained in Alameda County with his Pennsylvania-born wife, Henrietta, until his death in 1908. The land was productive. In 1867 the Farrellys raised a carrot which measured 36 inches long, 31 inches in circumference, weighing 31 pounds without the leaves. He also opened a fruit orchard in San Leandro. In 1863 he was elected as a county supervisor. In that role, he was chairman of the meeting when it adopted his memorial resolution after the assassination of President Lincoln. He became an officer in the Union Savings Bank of Oakland, founded in 1860. A ready contributor of large sums to charities, Farrelly was highly respected throughout the county. The inheritance tax from his estate was appraised at $19,394.[17]

William H. Zerbe was born August 18, 1824. At the mines in Nevada County he "came to the conclusion that the mines would not pay," but he kept at it, spending the winter at his trade as a carpenter. After nearly three years, he shipped for Peru as a captain's steward, carpentered there for four months, then sailed as a ship carpenter around Cape Horn and went home. He later became a wholesale liquor salesman in Reading.[18]

Uriah Green, who indicated he was twenty-nine years old when he applied for a passport in 1849, opened a mercantile business in San Francisco in partnership with a man named Riggs. As Allen indicates in a letter, Riggs was a gambler and cost them the store. Uriah then went mining in Calaveras County.

The schooner *Newton*, on which the Reading party had sailed, held thirty-seven passengers. Another group of gold seekers, from Philadelphia, composed the balance of the passenger list. Those men were: Moses Albright, Samuel M. Dane, Nicholas Davis, Horace B. Dick, Thomas Diehm, Jr., Patrick M. Foley, A. Hallman, Hobart Hare, Lewis Hiough, Dr. J. Lukens, Capt. David

M'Cowell, A. D. Marshall, Conrad Meyer, Edwin A. Rigg, Robert Robinson, Robert Scott, J. R. West, Daniel Wineland, and Anthony M. Zane.[19] Several of them are mentioned in the letters.

The eleven men who had left Reading prior to organization of the Reading California Association were: Franklin Bitting, Augustus Fisher, William Scharman, Charles Deem, Dalrymple Turnbull, Aeneas Dugeon, John Doughton, George W. Stillwell, John Stroup, W. C. Stebbens, William G. Bowman, William C. Leavenworth, and Joseph Fricker.[20]

$\mathcal{C}\!\!\sim\!\!\mathcal{O}$

APPENDIX B

The Recollection of Edwin A. Sherman

As described in the preface, a persistent claim that Allen and Hosea Grosh were daguerreotype artists has caused their names to appear in numerous authoritative lists of early photographers. We offer the following discussion to evaluate—and ultimately to discount—the assertion that the brothers were engaged in this endeavor.

In the opening decade of the 1900s, '49er Edwin A. Sherman wrote his lengthy memoirs, in which he included a story of how he guided a company of gold seekers across Mexico. Major portions relative to California were published in 1944.[1] Sherman used "notes and memory" for the writing, with annotation later provided by his son Allen.

A paragraph in Sherman's published recollection can be regarded either as an important piece of additional information about the Grosh brothers story, or as a misleading tale based on a false recollection. Sorting out the difference is a challenge, but the exercise is important because some secondary sources, based on the strength of Sherman's recollection, have listed the Grosh brothers as daguerreotype artists.[2]

In writing of leading the "Camargo company" of gold seekers across Mexico in 1849, Sherman included the following passage in his memoir:

[In the central Mexico city of San Luis Potosí we were] guests in peace, or prisoners of war, according to the conditions and circumstances. . . .

We were detained there ten days but lost nothing by it.

The inhabitants had not learned of the discovery of gold in California; and that we had crossed the Sierra Madre was a miracle to them. And strange as it may seem, no one in that city had ever heard of or seen a daguerreotype! Taking the two brothers, Hosea B. and Allen E. [*sic*] Grosh, daguerreotypists with me, I called on the governor and had them show him the daguerreotypes, taken in Tampico, of General La Vega and others, and told him that if he would sit for them, we would be very much pleased. He was delighted as well as surprised. The Groshes looked for a back room in which they could fix up a dark closet for their chemicals—quicksilver, bromide, and iodine for development—then back for their camera and fix-

ings. When they returned, the governor sat for his picture. It was successful from
the start, and when the daguerreotype was finished and put in a fine case, he was
in ecstasy, and was still more astonished when it was presented to him. He then
had daguerreotypes taken of his whole family, which he insisted on paying for.
The Groshes had such a run of business that their stock of plates and chemicals
was exhausted. At the drug stores they could procure only a limited supply of the
chemicals. The governor wanted to know what the plates were made of. On being
told that they were made of thin copper, plated with silver, he asked, "What is to
prevent more being made entirely of silver?" Being informed that there was nothing to prevent but the expense, he sent for the druggists in the city to supply what
chemicals they could and set the silversmiths to hammering and rolling out silver
plates, cutting them up and polishing them. The Grosh brothers were kept hard at
it for a whole week, until the chemicals were entirely exhausted and they had to
close business.[3]

Sherman wrote in some detail of his efforts in 1849, when he was twenty
years old and a veteran of the Mexican-American War, to organize three companies of gold seekers in Pennsylvania. Joining the final group, the Camargo
company, aboard the brig *Thomas Walter,* he ventured forth to seek his fortune
in California.[4] The brig sailed on February 1. The Reading California Association, in which Allen and Hosea Grosh were members, left Philadelphia for
Tampico four weeks later on February 28. They had sent their heavy cargo
ahead on a ship that would sail around Cape Horn, the southern tip of South
America.[5]

Benajah Jay Antrim of Sherman's Camargo company kept a journal and
sketchbooks while crossing Mexico, recording their arrival in San Luis Potosí
on April 22 and departure at 5:30 A.M., Sunday, March 25. He was planning a
series of paintings of Mexico to create a public exhibition, for which he made
watercolor paintings of the countryside. After trying his hand at mining in
Placerville in 1850, he is listed as a daguerreian photographer in Sierra County
in the Special California Census of 1852.[6] In 1851 Daniel Woods, also of the
Camargo company, wrote: "On the 22d of March we entered San Luis Potosí.
This is a large city, possessing considerable wealth. It is near the silver mines,
and contains a mint." After describing their visit, he echoed Antrim, writing:
"We spent two days in the city to give rest to our animals, and then proceeded
on our way toward Guadalajara."[7] Woods wrote that the Camargo company
reached Guadalajara, about 65 road miles west of Potosí, "on the 2d of April
1849."

Meanwhile, the Reading California Association had arrived at Tampico late
in March, where it was delayed for ten days. On April 9, the night before leaving Tampico, Allen wrote that "The party which came by the *Thomas Walter*

from Philadelphia, about a month before we did, were fleeced pretty hand-somely." He cited three-card monte as the "fleecing" event, but he did not describe how or from whom he gained that information. In the April 9 let-ter, Allen wrote that "General Urrea . . . commands this division . . . *in the absence of* General La Vega" (emphasis added). General De la Vega, whose daguerreian image Sherman said the brothers had taken, left Tampico with his troops between March 9 and April 1 to march on a band of dissidents in the mountains.[8]

In the June 13 letter, Hosea wrote that the Reading party left Tampico "on the 10th of April . . ." and gave their position on Tuesday, April 17: "There being a river [the Tamesi] to ford we did not intend to start till daylight." The river is far to the east of Potosí.

On April 2 the Camargo company of E. A. Sherman was already west of Guadalajara, some 265 miles west. Writing from San Francisco on December 1, Hosea recalled: "At San Luis Potosi the government seized the mules that we had hired . . . this detained us ten days." The Grosh brothers arrived in San Blas, on the Pacific Coast, the evening of June 23. Daniel Woods recorded the Camargo company's sailing from San Blas on April 12. There is no evidence that E. Allen Grosh, Hosea B. Grosh, and Edwin A. Sherman were in any Mex-ican cities, especially San Luis Potosí, at the same time.

The equipment required to take and develop daguerreian images includes a large camera and tripod, silver plates and a box for them, iodine and bromide boxes, a mercury cabinet, a leveling stand, a flat dish for washing, and a hand-buffing tool for polishing the plates. It all traveled in a large wooden trunk, the type of equipment the Reading company would have shipped "around the Horn."

In the first letter, dated February 27, 1849, from Philadelphia, the Grosh brothers wrote that they were not able to comply with a request "to have our miniatures taken here before we left." Nothing in any of the letters indicates that the brothers had the ability to take the images themselves or that they possessed the cumbersome box of equipment involved.

During the 1890s, two articles in the *San Francisco Call* identified the Grosh brothers as discoverers of the Comstock Lode and told the story of their deaths.[9] Eliot Lord's *Comstock Mining and Miners,* mentioning the Grosh brothers, had been published in 1883, the same year that Edwin Sherman left Nevada and became a mining consultant in Oakland. Then, in the opening decade of the 1900s, Sherman wrote his memoir. It is clear from the publica-tion that Sherman held himself in high regard and enjoyed demonstrating his relationships with well-known people, so it is easy to conclude that he came to believe he had a connection with the Grosh brothers of the newspaper articles. Despite Sherman's recollections of images having been taken, staff at the Archivo Historico de San Luis Potosí is not aware of any photographs of Julián de los Reyes, who became governor of the Mexican State of San Luis Potosí in early 1849.[10]

Considering the letters written by the brothers, as presented here, it seems unlikely that they were capturing photographic images during their trip across Mexico. Unfortunately, there may never be a clear explanation as to why Sherman created the tale of the Grosh brothers as daguerreian photographers.

APPENDIX C

Financial Matters and the Subsequent Lawsuit

By virtue of prospecting in the western Great Basin, the story of the Grosh brothers became entangled with one of the greatest episodes in the history of mining, namely the discovery and extraction of the Comstock Lode. As mentioned in the preface, the ensuing legal contest about the validity of a Grosh claim to the Comstock likely played a role in the preservation of the letters.[1] That beneficial legacy of an assertion of Comstock ownership aside, it is appropriate to offer some evaluation of the idea that the Grosh family had a defensible claim to some of the proceeds from the businesses that mined in the district. Because the story of litigation involving the discoveries of the Grosh brothers postdates their letters and deaths, it is ancillary to the central theme of this book. There are, however, documents on file in the Grosh Collection at the Nevada Historical Society dealing with the legal aftermath following the deaths of the brothers, and these warrant a brief discussion.

The story of the brothers did not end with the burials of Hosea Ballou and Ethan Allen Grosh. Normally, Allen would have opened probate for his brother's estate in the court at Genoa. But Judge Orson Hyde had suffered severe frostbite during the previous winter, and after a long convalescence he returned to Salt Lake City. There were no probate filings in the Carson County Probate Court between June 1856 and September 1859.[2]

In the August 1857 letter before Hosea's death, Allen discussed their plan to return to California for the winter. That trip cost Allen his life. There is no record of probate being opened for Allen's estate in Placer County, California, where he died, and as a result of this, neither of the brothers' estates was resolved through the legal system.

The Comstock Mining District was founded in 1859, less than two years after the brothers died. During the following twenty years the mines produced hundreds of millions of dollars in silver and gold. There was ample evidence that the Grosh brothers had identified the existence of silver in the region before the strikes of 1859, so it was easy for potential heirs to imagine that some wrong had been done. In gathering every scrap of legal evidence, the family assembled the letters from the brothers and saved them together with other documents and relevant recollections. The material related to the legal turmoil that followed their deaths now comprises the Grosh Collection archived at the Nevada Historical Society in Reno.[3]

There were two issues related to a legal challenge to the ownership of the Comstock Mining District. The first concerned whether the brothers had discovered the ore body that was producing the profits or they were merely some of the first to realize silver deposits existed in the region. Second, there was a question as to whether any sort of ownership of any ore body could remain valid after two years of neglect. After all, whatever they found was at most only preliminarily developed and then abandoned by virtue of their deaths.

Eliot Lord, in his definitive history, *Comstock Mining and Miners* (1883), placed the Grosh challenge to establish ownership of the Comstock with other frivolous examples of litigation. Beginning in 1863, the Grosche [*sic*] Gold and Silver Mining Company began the process of suing the Gould and Curry and the Ophir mining companies in the Twelfth District Court in California. As Lord pointed out, "Whether the Grosh brothers ever made a location on the side of Mt. Davidson is immaterial. It is certain that they developed no ledge and that all their record notices were lost. To found a claim to 3,750 feet of the Comstock ledge upon their vague discoveries was simply preposterous; yet the Grosche Gold and Silver Mining Company was incorporated in 1863 with a nominal capital of $5,000,000, afterward increased to $10,000,000, in order to maintain this claim, which embraced the richest portion of the ledge."[4] Lord further noted that the Gould and Curry alone spent $12,933.30 defending against the legal contest, which was not dismissed until March 9, 1865: the company "had been forced to assemble witnesses from all parts of the country, and fortify their title against an assault which was a clear case of black-mail."[5]

Myron Angel's *History of Nevada, 1881,* includes a statement by Mrs. Laura M. Dettenreider, who knew the brothers while they worked in Gold Canyon. She became one of the claimants in the lawsuit, because she maintained that the young men had verbally promised her a share in their mine. This is followed by the statement of Reverend A. B. Grosh, their father. In both cases, dates and places are confused. For example, Dettenreider may have misunderstood references to the Grosh claims at Sugar Loaf in the California gold country with the Sugar Loaf of Six Mile Canyon, extending east from present-day Virginia City. Such confusion would have considerable consequences, since, had the brothers been prospecting above Sugar Loaf near Gold Canyon, Utah Territory, their chances of finding the Comstock Lode would have been better. It is more likely that they remained in the American Ravine, an offshoot of Gold Canyon near present-day Silver City, far below the valuable core of the Comstock Lode.[6]

On March 26, 1853, Allen wrote that they had information about good prospects near the summit of the Sierra, which would have them investigating to the west of the Virginia Range and its Comstock Lode. In addition, in the letter dated November 22, 1856, Allen made it clear that Frank Antonio, the man known as "Old Frank," inspired them, in 1853, to go to Utah Territory.[7]

The brothers arrived in Carson Valley, Utah Territory, on July 30, 1853, according to a letter written the following day from Mormon Station, later called Genoa. They then went back into the Sierra and prospected south of Lake Tahoe, apparently returning to Carson Valley in October of that year. On December 3, 1853, Hosea authored a letter from Gold Canyon, the first indication that the brothers had arrived in the area subsequently known as the Comstock Mining District. What followed was the longest gap in the sequence of letters, with the next correspondence from Allen dated November 8, 1854, and sent from Sugar Loaf, El Dorado County. The letter hints that they had just arrived back in California, suggesting that they spent the winter and following summer in Utah Territory.

There is no further mention of returning to Utah Territory until September 18, 1856, when Allen writes that they were leaving that day for their silver mine. A letter dated November 3, 1856, indicates that they had returned to California "three days ago," having found "two veins of silver at the forks of Gold Canon." On November 22, the brothers suggest that they hoped to return and winter in Utah Territory. Snow, however, curtailed their ambitions, and they did not return to Gold Canyon again until May 27, 1857.

Dettenreider maintained that she met the Grosh brothers in 1854, and she correctly recalled that they spent the winter of 1854–55 in California. She then asserts, however, that the brothers came back to Utah Territory in 1855. There is no indication that this occurred in any of several letters dated from that year. It is more likely that Dettenreider was recalling the brief visit in September 1856, and again, the difference is critical to the idea that the brothers had discovered the Comstock Lode. Dettenreider asserted that "Hosea said that they were hurrying away because they had to reach the Sugar Loaf in Six-mile Cañon that night, where they proposed making a camp at a spring."[8] She then went on to indicate that the brothers "intended to prospect farther for silver in the vicinity of where they had found it the year before," and she added that the brothers promised her a share of the mine. In fact, after their brief 1856 visit, the brothers hurried back to Sugar Loaf in California, and it is possible that Dettenreider misunderstood their comment, because Sugar Loaf is also the name of a prominent landmark in Six Mile Canyon, directly below what would become Virginia City. It was also, of course, in her best interest to recall events in this way, since her memory placed the brothers in Six Mile Canyon, looking for silver above, and more likely, therefore, to have found the actual Comstock Lode.

Dettenreider then recalled that she returned to Gold Canyon in 1857 and happened upon the brothers in their cabin at the American Flat Wash near a fork in the canyon below present-day Silver City, the exact location where most evidence indicates they worked. Tragically, it appears she found them just as Hosea was nursing the wound that would prove his undoing, so it seems clear that she witnessed a pivotal moment. But her memory was flawed and perhaps self-serving.[9]

Allen describes a "small ravine to the right fork of the main canon" on September 7, 1857. Given the context, this can only be taken to mean the location of the previous investigations in the American Ravine, otherwise known as the American Flat Wash. Reverend Grosh asserts, however, that references to this place should be taken to mean "Gold Hill," which is perched high above and several miles to the north from the location where the brothers were working. The distinction is significant, since it is the difference between an ore body worth tens of millions and one worth far less.[10]

The fact that Henry Comstock came to be associated with this story added credence to the idea that the Grosh brothers had discovered the Comstock Lode and that Henry Comstock had stolen, or "jumped," the claim. By implication, Reverend Grosh and Dettenreider would have been entitled to some of the proceeds of the corporations that worked the Comstock Lode. Reverend Grosh asserted, in his letter published by Angel:

> . . . if reports may be relied on, Comstock himself told so many differing stories in accounting for his possession and sale of the lode, that it came to be believed that he took possession of books, maps, and other papers which Allen had boxed up for safe keeping, and thus learned of the existence of the mines they had discovered, and claimed—sometimes as his own discovery; sometimes as having been left in his charge, for which he was to receive one-third or one-fourth; sometimes as their partner; and sometimes as being on the spot, and therefore nearer to them than any distant heirs: having the best right, that of possession.[11]

While Henry Comstock was a scoundrel who was capable of talking himself into shares of the profitable claims of others, it does not follow that the Grosh brothers had made a major discovery or that Comstock stole it. Comstock's strategy of maneuvering his way into claims clearly paid off in 1859. If that was his intention with the legacy of the Grosh brothers—and it is not clear that this was the case—Comstock would have been pursuing a dead end, since there is no good evidence that the Grosh brothers found the actual lode or any ore body of significance.[12]

With hundreds of millions of dollars at stake, and a lingering belief that the unfortunate Grosh brothers were the first to lay claim to the Comstock Lode, it is no surprise that the threat of legal action emerged. The heirs of the Grosh brothers hired noted lawyer Benjamin F. Butler to litigate their assertion of ownership. Prominent Nevada journalist and newspaper editor Sam P. Davis indicated that Butler conducted a thorough examination of the claim and concluded that a good argument could be made that the heirs had a right to demand compensation in view of a reasonable assertion of at least partial ownership of the Comstock. Nevertheless, Butler added that "the defendants were men so thoroughly entrenched in possession, and having unlimited money at their command that they would be able to buy up any jury that could be selected to try the case, and that, under the circumstances, the winning of such a case would be an impossibility."[13] Whether or not litigation was warranted, the Grosh heirs subsequently dropped the legal action. The various legal contests, the evidence, and the positions of the parties involved could be the subject of an in-depth study, but little is to be gained here by elaborating further. What is important to point out in the context of these letters is that the legal argument that the brothers had discovered the Comstock Lode— regardless of reasonable questions about the story's validity—continued to echo in Comstock histories for decades.

The Grosh brothers were clearly working at the site of an outcropping in the vicinity of Silver City, removed from the Comstock Lode proper. In spite of the humble nature of their claim, their observations of Gold Canyon in the 1850s are extremely valuable in themselves. Besides that, the place of the Grosh brothers in Nevada history stems not from an alleged discovery of the Comstock Lode itself, but rather from their early recognition of the existence of silver in the region, a fact recognized previously by others but never with the notoriety of Allen and Hosea.[14]

$$\mathcal{C}\!\!\sim\!\!\mathcal{G}$$

NOTES

Many of the people mentioned in the Grosh letters appear in the various decennial United States Census manuscript records. In the following notes, simple references acknowledge the Census and the year in which information appears.

INTRODUCTION

1. Readers who are vague about the period may misunderstand references to prospecting in Utah: the western Great Basin was part of Utah Territory until the 1861 organization of the Nevada Territory.

2. Alexander Harris, *Biographical History of Lancaster County* (Lancaster, PA: Elias Barr, 1872). In 1823 Jacob Grosh became a prime suspect in a conspiracy to break into a local bank to destroy mortgage documents. The case was never brought to trial, and for some, Jacob was a hero standing against the tyranny of corrupt banking practices. F. Lyman Windolph, "A Mysterious Bank Robbery," *Pennsylvania Magazine of History and Biography* 94, no. 3 (1970): 384–95.

3. Franklin Ellis and Samuel Evans, *History of Lancaster County, Pennsylvania, with Biographical Sketches of Many of Its Pioneers* (Philadelphia: Everts and Peck, 1883), 630. C. C. P. Grosh became a justice of the peace in the county on April 12, 1864 (ibid., 628).

4. After the deaths of his sons, Aaron Grosh accepted an appointment in the Department of Agriculture in Washington, DC, as its first librarian from 1867 to 1869. He then cofounded the National Grange of the Order of Patrons of Husbandry in late 1867. Often called simply "the Grange," the organization continues to serve as an economic and political union for farmers and their families. Grosh also authored a book intended to help members of the Grange to understand the organization, its teachings, and its benefits.

5. Passport applications, http://www.ancestry.com, accessed February 28, 2011.

6. *Christian Ambassador* (Auburn, NY), February 20, 1858. This Universalist newspaper commenced January 1851; Reverend Thomas Jefferson Sawyer was the editor in 1857.

7. Ibid., November 14, 1857.

8. Charles Howard Shinn, *The Story of the Mine: As Illustrated by the Great Comstock Lode of Nevada* (1910; reprint, Reno: University of Nevada Press, 1980), 27.

9. Among the many volumes written on the California Gold Rush, Holliday remains one of the more highly regarded. J. S. Holliday, *The World Rushed In: The California Gold Rush Experience* (New York: Simon and Schuster, 1981) and *Rush for Riches: Gold Fever and*

the Making of California (Berkeley and Los Angeles: University of California Press, 1999). For an additional volume of note, see Malcolm J. Rohrbough, *Days of Gold: The California Gold Rush and the American Nation* (Berkeley and Los Angeles: University of California Press, 1997).

10. Sally Zanjani, *Devils Will Reign: How Nevada Began* (Reno: University of Nevada Press, 2006), 46.

11. Rohrbough, *Days of Gold,* discusses many of the aspects of California and the Gold Rush. Regarding failure, see, for example, 256–66, but Rohrbough discusses many themes that manifest in the Grosh letters.

12. Anton Phillip Sohn, *Healers of Nineteenth-Century Nevada* (Reno: Greasewood Press, 1997), 73.

13. In 1961 the Universalist Church merged with the American Unitarian Association to become the Unitarian Universalist Association. David E. Bumbaugh, *Unitarian Universalism: A Narrative History* (Chicago: Meadville Lombard, 2001).

14. Zanjani, *Devils Will Reign,* 42, 44, 46; Shinn, *The Story of the Mine,* 29.

15. Mark Twain, *Roughing It,* edited by Harriet Elinor Smith and Edgar Marquess Branch et al. (1872; reprint, Berkeley and Los Angeles: University of California Press, 1993), 357. See also Eliot Lord, *Comstock Mining and Miners,* Monograph of the U.S. Geological Survey, vol. 45 (Washington, DC: Government Printing Office and Department of the Interior, 1883), 57; and George Lyman, who, in a citation for his historical novel, *The Saga of the Comstock Lode* (1934; reprint, New York: Charles Scribner's Sons, 1951), 361, 3:4, quotes the Spanish proverb as "Para trabajar una mina de plata se necesita una mina de oro."

16. Zanjani, *Devils Will Reign,* 42–44.

17. F. W. McQuiston, *Gold: The Saga of the Empire Mine, 1850–1956* (Grass Valley, CA: Empire Mine Park Association, 1986).

18. For example, John D. Mitchell, *Lost Mines of the Great Southwest* (1933; reprint, Glorieta, NM: Rio Grande Press, 1990); and Thomas E. Glover, *The Lost Dutchman Mine of Jacob Waltz* (Phoenix: Cowboy Miner Productions, 2000).

19. Wright describes seeing the early works in 1860 in his free-flowing quasi history: Dan De Quille [William Wright], *The Big Bonanza* (1876; reprint, New York: Alfred A. Knopf, 1947), 14–16.

20. The map is entitled "Washoe Silver Region" and is available online at http://contentdm.library.unr.edu/cdm4/item_viewer.php?CISOROOT=/hmaps&CISOPTR=4781&CISOBOX=1&REC=3, accessed January 17, 2011; *Sacramento Daily Union,* August 6, 1863; the DeGroot articles were republished as Henry DeGroot, *The Comstock Papers* (Reno: Grace Dangberg Foundation, 1985), 5–6; Wright, *The Big Bonanza;* Lord, *Comstock Mining and Miners;* Hubert Howe Bancroft, *Chronicles of the Builders of the Commonwealth* (San Francisco: History Company, 1892), 4:190; Shinn, *The Story of the Mine,* 27–30, 32.

21. Lyman, *Saga of the Comstock Lode,* 16–31. For the historical marker, see Erwin G. Gudde, *California Gold Camps* (Berkeley and Los Angeles: University of California Press, 1975).

1849

1. Flack's name appears as "Flag" in the article. Other spelling has been modernized. Although many traveled as individuals or in small, informal groups, the organization of companies, business ventures linked by formal agreement and financial support, was also common. The overland route through central Mexico, chosen by the Reading California Association, was one of many alternatives for those in the eastern United States. Other routes included passing over the Panama Isthmus, crossing North America, and rounding the Horn of South America. Holliday, *The World Rushed In* and *Rush for Riches;* and see Rohrbough, *Days of Gold,* 55–66 (for all, see introduction, n. 9).

2. The writer may be indicating that this letter was written at noon. A corner of the page is torn off immediately after the "12" where the word "noon" may have been.

3. Cape May, New Jersey, and Cape Henlopen, Delaware, both at the mouth of the Delaware River.

4. The poem is "The Sea" by Barry Cornwall [Bryan Waller Procter] (1787–1874).

5. *Levant,* a 382-ton ship launched in Newbury, Massachusetts, in 1831, sailed from Philadelphia on February 26, 1849. Charles W. Haskins, *The Argonauts of California* (New York: Fords, Howard, and Hulbert, 1890); *Libera Martina Spinazze's Index to the Argonauts of California* (New Orleans: Polyanthos, 1975). Originally compiled under the direction of the Society of California Pioneers and published with the permission of the California State Society, National Society of the Daughters of the American Revolution, Spinazze's volume is an alphabetical index to names in Haskins, where the last 141 pages include a compiled list of the "Names of Pioneers Who Came by Land and Sea to California in 1849." These lists show the names of passengers on specific ships, such as that used by the Grosh brothers, a format not included in Spinazze. *Special Lists #22 [Alphabetical] List of American-Flag Merchant Vessels That Received Certificates of Enrollment or Registry at the Port of New York, 1789–1867,* 2 vols. (Washington, DC: National Archives; National Archives and Records Service, General Services Administration, 1968).

6. These people are as follows: "Conrad Meyer, Horace B. Dicke and Captain J. R. West." "Off For California," *New York Herald,* March 6, 1849. They were among the nineteen passengers aboard with the eighteen-member Reading California Association. West was in Captain J. Boyd's cavalry unit, commanding officer of a four-officer, seventy-six-trooper unit. He was active in local politics in San Francisco, where he was promoted to major in the California Volunteers. *Alta,* June 13, 1854.

7. *Thomas Walter* was a brig, displacing 149 tons and built in Wilmington, Delaware, in 1846. *Special Lists #22.* Among the fifty passengers were the forty men of the Camargo company from Philadelphia. The story of that company's passage across Mexico is related in Daniel B. Woods, *Sixteen Months in the Gold Diggings* (New York: Harper, 1851), and in a reminiscence, Allen B. Sherman and Edwin A. Sherman, "Sherman Was There: The Recollections of Major Edwin A. Sherman," *California Historical Society Quarterly* 23, no. 3 (1944). A portion of that article is reproduced in appendix B.

8. Words appear at this point, but they have been crossed over. The lined-out text reads as follows: "But I expect they were a pretty rough party, and deserved what they got if not more."

9. General Don José Cosme de Urrea and General Rómulo Díaz de la Vega.

10. In the United States a league is generally accepted as 3 statute miles in length. It is not a world standard, often differing from nation to nation.

11. Also known as the Guyalejo River, a tributary of Pánuco emptying into the Gulf of Mexico at Tampico.

12. Two other attempts at spelling "diarrhea" were scratched out.

13. This archaic term refers to a succession of fevers and chills, particularly associated with malaria. Part of the word here is missing, but the "a" and part of the "g" are present, and the brothers use the term "ague" elsewhere in this context.

14. All that remained here was apparently the end of the word, which can be reconstructed as "meat" or "fat." The page at this point is badly scorched, and the editors have relied here and at other points on family transcripts for words that are no longer visible.

15. A passage relative to the brothers' reaction to the crossing of Mexico is quoted in Lord, *Comstock Mining and Miners*, 25 (see introduction, n. 15). Dr. Richard Maurice Bucke, who played a pivotal role at the end of the Grosh story, apparently gave Lord access to the now-lost journals of the Grosh brothers. Based on this source, Lord quoted the brothers' reference to the trans-Mexico journey as being across "a barren country with few trees and these almost leafless, stampedes and straying mules and horses, poor provisions, insults of all kinds day and night, bad roads and in places no roads, and attacks of malarious fever and dysentery." Woods, *Sixteen Months*, 20-39, describes a route similar to that of the Reading company.

16. Dr. John H. Behne, physician, Reading, Pennsylvania, born in Prussia, ca. 1800 (Census, 1850 and 1870).

17. This letter is not among the documents in the Grosh Collection at the Nevada Historical Society. The extracts were contained in a long article in the *Reading Eagle*, Sunday, August 11, 1929. "Reading Had 29 in the Gold Rush to Pacific in '49; Endured Terrible Hardship," http://news.google.com/newspapers?id=YIohAAAAIBAJ&sjid=oZcFAAAA IBAJ&pg=1795,1926620&dq=california-association&hl=en, accessed February 27, 2011.

18. An area in the State of Jalisco, Mexico, southwest of Tepic, called the Plan de Barrancas (Plain of Gorges).

19. The use of the term "company" throughout the letters is ambiguous. Although it may sometimes simply mean "cohort," it is clear that at other times it is used to indicate a group of people linked by an official agreement to pursue a commercial enterprise. Context assists in determining the meaning of the word, but sometimes the precise meaning cannot be ascertained.

20. The idea that the captain would "find us" indicates that he would provide meals. The brig *Olga*, Captain John C. Bull, was in San Francisco Bay in August 1849. When the crew abandoned the ship for the Gold Rush, Bull went home to Boston. He eventually returned with his family and settled in Humboldt County, California. Leigh H. Irvine, *History of Humboldt County California: With Biographical Sketches of the Leading Men and Women of the County* (1915; reprint, Provo, Utah: Generations Network, 2005), 221. In some ships, those in second class were called steerage passengers. The steerage area of the ship was once used to accommodate those traveling on the cheapest class of ticket

and offered only the basic amenities, typically with limited toilet use, no privacy, and poor food. William Henry Smyth and Edward Belcher, *The Sailor's Word Book: An Alphabetical Digest of Nautical Terms* (London: Blackie and Son, 1867), 654.

21. Alpheus Bull (no known relation to Captain John C. Bull of the brig *Olga*) was born at Bullville, Long Island, New York, on June 16, 1816. He was the fifth of fourteen children born to Henry and Jane Stitt Bull. Little is known of his early life except that he was an itinerant preacher of the Universalist Church who spent his time in the East traveling, preaching, and doing church-related work. In March 1849 Alpheus Bull joined the Lafayette California Mercantile Mining Company, which was organized for an overland trip to California. The Gold Rush was well under way, and although Alpheus Bull was not convinced of the abundance of gold in California, he anticipated using his ministerial calling during an exciting trip. The party left Lafayette, Indiana, and traveled through Mexico to Mazatlán and then by ship to San Francisco, arriving in late July 1849. Alpheus and several others were successful prospecting at Nigger Hill. (This name has disappeared, but research shows that Murphy's Flat is on the side of Nigger Hill and Watson Gulch, nearby.) Within the year Alpheus Bull, George Baker, and William Robbins decided to become merchants of mining supplies. Bull later became involved with the Comstock Lode mines and with the marking of the Grosh brothers' graves. He died in 1890. http://cagenweb.com/shasta/bios/bullalpheus.html, accessed January 31, 2011.

22. Meaning the 13th of September 1849.

23. "No. 6" was a codeine-laced high-alcohol liquid that was heated, and the vapors inhaled, to treat asthma. What the "injection" mentioned in the letters involved is not clear. Frontier medicine is discussed in George W. Groh, *Gold Fever, Being a True Account, Both Horrifying and Hilarious of the Art of Healing (So-Called) During the California Gold Rush* (New York: William Morrow, 1966), and in the autobiographical volume Dr. Reuben Know, *A Medic Fortyniner* ([Verona, VA:] McClure Printing, 1874).

24. Dr. May, a physician on Pike Street, San Francisco (Charles P. Kimball, San Francisco City Directory [San Francisco: City of Commerce Press, September 1850]), whose wife, Susan, joined him in 1851, survived the disease and moved from San Francisco to Sacramento, where he opened the Boston Drug Store (Sacramento Transcript, October 16, 1850) during the cholera outbreak in that city. William Roy Shurtleff, *The Miller and Simmons Families Genealogy*, vol. 2 (Lafayette, CA: Pine Hill Press, 1993), 196–97. Eclectic medicine relied on physical therapy combined with medicines derived from plants.

25. William H. Zerbe, born ca. 1825 in Pennsylvania, a bartender, died February 11, 1894, in Reading, Pennsylvania (Census, 1870 and 1880; http://www.ancestry.com, accessed November 20, 2010). Zerbe's journal, begun in July 1850 at Rough and Ready, California, continued until he boarded a ship to return to Reading that November. See the William H. Zerbe Diary, Department of Special Collections, Hesburgh Libraries of Notre Dame University, Indiana.

26. Although Axe was in camp by the end of November, the American barque *Hortensia* was not certificated in San Francisco until May 9, 1850 (William Heath Davis Papers, BANC MSS C-B445, Bancroft Library, University of California, Berkeley). The barque appears to have been a slaver out of Baltimore when first constructed in the 1830s and

then an emigrant ship from Europe. Brockman Family, Port of New Orleans Record Group 36, U.S. Customs Service, Microfilm roll 12, National Archives and Records Administration, Washington, DC.

27. The use of mercury or "quicksilver" in retrieving gold from placer sands or from crushed ore was the standard approach at the time. Gold and mercury form an amalgam that could be roasted in a cast-iron pan, driving off the mercury and leaving the precious metal. Of course, the fumes are toxic. The process could also be used with silver, but silver ore usually presented additional chemical complications.

28. The actual specific gravity of gold is 19.32.

29. This was probably Taylor Nichols, born in 1803, a blacksmith (Census, 1850). Cornwall was a highly respected international center of mining at the time.

30. "Burning lime" refers to the process of heating limestone to drive off the carbon dioxide, creating "quick lime," which was mixed with water and sand to produce a mortar.

31. John Lash was a machinist in Reading, living with his mother and siblings (Census, 1850).

32. Horace Greeley wrote in his *New York Tribune* that the radical Democrats of New York would "burn down their own barn to get rid of a rat infestation." The Barnburners, mentioned again in a letter dated September 18, 1856, stood in opposition to the Hunkers. See chapter 4, note 7, for a discussion of Hunkerism.

33. Daniel Wineland of Easton, Pennsylvania, born in 1828, was a rope maker who was on the *Newton* but not a member of the Reading company, which included the Grosh brothers (Census, 1870).

34. Many California gold-country locations employ the term "bar," as in "Michigan Bar" near Sacramento. A bar is a low-level deposit of alluvial material (sand and gravel) in a stream or river formed by the gradual addition of new material. Bars typically form along the short or inside radius of curves above or below the waterline of streams. Bars may also form where the current slackens or changes direction. Also referred to as "gravel bars" or "sandbars," these were excellent locations to find gold.

35. The *Pacific News* was published three times a week from August 25, 1849, to March 1, 1850. At the time of this letter, in May 1850, it appeared six days a week. The newspaper went out of business in 1851.

36. Nathaniel Davis was among the nineteen passengers on the schooner *Newton* from Philadelphia to Tampico in 1849 who were not members of the Reading California Association.

37. Lewis Briner, a dry-goods merchant in Reading. The extensive Ritter family had settled in Berks County in 1753. Family patriarch Christian Ritter, seventy in 1849, and Briner, thirty-three, were both 1832 charter members of the First Universalist Church of Reading. They won their struggle against Reverend A. B. Grosh, who had become the society's fourth minister in 1845. He was replaced in early 1850 by Reverend J. Shrigley. By March 1850, Reverend Grosh had accepted a call by a congregation in New York State. Morton L. Montgomery, *History of Berks County in Pennsylvania* (Philadelphia: Everts, Peck, and Richards, 1886), 790; Census, 1850.

38. Congressman (i.e., M.C., or member of Congress) John Freedley was born in Norristown, Pennsylvania, on May 22, 1793. He attended public schools and Norristown Academy and served as an assistant to his father, who operated a brickyard. He studied law, was admitted to the bar in 1820, and commenced practice in Norristown. Freedley was elected as a Whig to the Thirtieth and Thirty-First Congresses (March 4, 1847–March 3, 1851) and died in Norristown, Montgomery County, Pennsylvania, on December 8, 1851. Charles Lanman, *Dictionary of the United States Congress* (Washington, DC: Government Printing Office, 1866), 139.

39. Congressmen and senators were paid eight dollars per day when in session from 1817 until 1856 (3 *Stat.* 404, approved January 22, 1818). The law was retroactive to 1817, and then it was changed to three thousand dollars per annum by 11 Stat. 48, approved August 16, 1856. Library of Congress, *Congressional Research Service Report*, 2008.

40. Point Concepcion is on the coast in Santa Barbara County, California. Allen is concerned for the steamship *Oregon*, which arrived in San Francisco the day he was writing the letter, Saturday, December 1, 1849. It brought "a large mail and 414 passengers. It is thought that her mail will be sorted and letters for this point [Sacramento] received by Tuesday next—ten days after her arrival!" *Placer Times* (Sacramento), December 8, 1849. There is no mention of accidents or delays in any of the contemporaneous newspapers. See also Raymond A. Rydell, "The Cape Horn Route to California, 1849," *Pacific Historical Review* 17, no. 2 (1948): 149–63.

41. John C. Frémont (1813–90) was an officer of the United States Army, an explorer, and a politician. He was the first candidate for the presidency of the newly founded, antislavery Republican Party, which nominated Frémont in 1856. In addition, he was the third military governor (1847) of California after the United States took control, a U.S. senator representing California (1850–51), and a territorial governor of Arizona (1878–81). The flamboyant Frémont had a checkered but remarkable career, the details of which are too great to list here. He was a favorite for many from California.

42. Mungo Park (1771–1806) was a famed Scottish explorer of Africa. His detailed travelogue, *Travels in the Interior Districts of Africa* (1799), was extremely popular. "Gulliver" is a reference to the fictional explorer of fantastic lands based on *Travels into Several Remote Nations of the World*, popularly known as *Gulliver's Travels*, by Jonathan Swift (1667–1745).

43. A gum blanket was rubberized to provide more protection against moisture.

44. The San Francisco fire of December 24, 1849, is described in Frank Soule, John H. Gihon, and James Nisbet, *Annals of San Francisco* (New York: Appleton, 1855), 241: "This morning about six o'clock, the awful cry of fire was raised in the city, and in a few hours property valued at more than a million dollars was totally destroyed. The fire began in Dennison's Exchange, about the middle of the eastern side of the Plaza, and, spreading both ways, consumed nearly all that side of the square, and the whole line of buildings on the south side of Washington Street between Montgomery and Kearny streets. . . . Scarcely were the ashes cold when preparations were made to erect new buildings on the old sites, and within a few weeks the place was covered as densely as before with houses of every kind."

45. A fire on December 23, 1849, destroyed a large part of the Stockton business district, which at the time was dominated by tents. Loss was estimated at two hundred thousand dollars. V. Covert Martin, *Stockton Album Through the Years* (Stockton, CA: Simard Printing, 1959).

46. On January 25, 1849, the first party of eleven gold seekers from the Reading, Berks County, Pennsylvania, area set sail for Nicaragua, a month ahead of the Reading California Association. The Reading men had joined the 150-member Gordon's California Association, composed mostly of young men from New York and Philadelphia. John Doughton was one of the Reading men, apparently a leader (see appendix A). *Reading Eagle*, August 11, 1929.

 1850

1. "Sandwich Islands" was the name given to the Hawaiian Islands by Captain James Cook in 1778.

2. When "going to see the elephant," using the phrase of the day for going to California, a gold seeker looked toward "making his pile," the then-colloquial phrase for becoming wealthy. In 1853 the *Sacramento Daily Union* published a booklet of woodcuts by Charles Nahl, with text by Alonzo Delano ("Old Block"), titled *The Miner's Progress; or, Scenes in the Life of a California Miner,* with the further subtitle "Being a Series of Humorous Illustrations of the Ups and Downs of a Gold Digger in Pursuit of his 'Pile.'" Hosea refers to "seeing the elephant" in the letter of June 13, 1849. Finding one's pile did not have to be a matter of finding gold: Jared Brown never found time to look for gold. He made his "pile" by plying his trade as a blacksmith in Coloma. In a letter home, he told his father that he sometimes made seventy dollars a day, an impressive wage even in those inflationary times. Jared Brown letter of August 11, 1851, California State Library.

3. *Susan G. Owens,* a clipper ship, was built in Baltimore in 1848, displacing 730 tons. After a 152-day voyage from Philadelphia and around the Horn, it arrived in San Francisco on October 12, 1849 (*Special Lists #22;* http://blog.cogswell.edu/2010/04 /the-history-of-cogswell-college/, accessed February 11, 2012). The *Owens* was in the Delaware River, thirty-five miles below Philadelphia, on May 7, 1849 (*Philadelphia Sun,* May 7, 1849). The Reading California Association apparently sent excess baggage to California on the ship.

4. Jackson's Creek is the area around today's Jackson, California, on Highway 88. In 1850 it was on the main road from Sacramento to the northern mines.

5. Word unclear; "addressees" is the best guess.

6. To "dance a hornpipe" was slang for being hanged, the victim's legs flailing about as though dancing.

7. Reverend Grosh was the second minister for the Society of Universalists in Utica, founded in 1834, where he served prior to the ministerial call to Reading in 1845. Thomas Griffith returned to Utica and in 1861 volunteered in the Third Regiment, New York Cavalry, seeing action in several Civil War battles. Ed Blake left New York on the ship *J. G. Costar* on February 21, 1849. Danby's son is probably Clarence Churchill Danby, twenty-two, who returned to Utica to practice law. Samuel W. Durant, *History of Oneida County,*

New York (Philadelphia: Everts and Fariss, 1878), 429, 648; National Park Service Online Civil War Soldiers and Sailors System, http://www.itd.nps.gov/cwss/, accessed March 22, 2011; Haskins, *The Argonauts of California* and *Spinazze's Index* (for both, see chap. 1, n. 5); Census, 1860.

8. Sending a lock of hair of the deceased as a remembrance was a common practice at the time.

9. On January 8, 1850, Sacramento endured major flooding of the waterfront. The event inspired discussion of constructing levees.

10. Hosea had apparently heard a fairly straight version of the incident. In Calaveras County a group of Chileans (many of them held in slavery) used force of arms to eject miners from profitable sites. At one point some non-Chileans were taken prisoner. While being marched away, they managed to free themselves from their bonds and relieve the sleeping Chilean guards of their weapons. The guards were then taken prisoner. The Stockton Rangers arrived to take over the Chilean prisoners. For a firsthand account of the little-reported incident, see "The Chilean War of Calaveras County," http://nevada-outback-gems.com/gold_rush_tales/california_gold_rush1.htm, accessed December 2, 2010.

11. "Dearness" meaning "high cost."

12. During the California Gold Rush, one-cent pieces were rarely used. Initially, commerce relied on the Mexican one-real coin, which was worth twelve and a half cents (the proverbial "one bit"). New Englanders often referred to these as shillings. Californians also used the half-real (also known as a medio or picayune), worth six and a quarter cents. As American coins became more common in the mining West, most transactions were done minimally with a five-cent piece, and pennies continued to be shunned for several decades, in part because the coin was so large, but also because it had so little value. In addition to minted coins, gold dust and barter frequently served as the basis for economic exchange. Fred N. Holabird, *Coins of the Comstock: The Pioneer Minor Coinage of Virginia City and the Comstock* (Reno: Sierra Nevada Press, 2009), 1-6.

13. When a printer's hand-set type was jumbled up, it was called "pi" and tossed into the "hell box," often left for an apprentice to sort out.

14. Colonel John "Jack" Coffee Hays (1817-73) was widely known in the 1850s for his actions as a member of the Texas Rangers. In 1842 Hays, then a captain, led a force of Rangers against the invasion from Mexico. He continued in a leadership role when the Rangers participated in the Mexican-American War (1846-48). His reputation for having given the Rangers a cohesive, disciplined group mentality preceded him when he relocated in California. He won election as sheriff of San Francisco, and then in 1853 was named California surveyor general. See James K. Greer, *Colonel Jack Hays* (New York: E. P. Dutton, 1952). Hays, who was thirty-three at the time of the election, was running as an independent and exploiting his reputation as a Texas Ranger.

15. Flowers of sulfur (which Allen spelled "sulpher"), also called precipitated sulfur or sublimed sulfur, is a powder that has long been used in folk medicines.

16. Jonas Winchester, traveling in 1849 on the ship *Tarolinta*, came to California from Marcellus, New York, 70 miles west of Utica; Haskins, *The Argonauts of California*, 448.

The *Pacific News* was published bimonthly in San Francisco from September 1, 1849, to October 30, 1850. In April 1850, the month of this letter, Winchester was appointed state printer (California State Library).

17. Colonel Ephraim N. Banks, an attorney living in Lewistown, Pennsylvania, was an examiner of cadets at West Point (Census, 1850); *Military and Naval Magazine of the United States* 1, no. 1 (1833).

18. John White Geary (1819–73) was appointed San Francisco postmaster in January 1849 and that same year was elected as the last alcalde of California. When the U.S. military took over California, it left the Mexican system of alcaldes, the municipal magistrates, intact. On May 1, 1850, as both the Democratic and Independent candidate, Geary beat Charles J. Brenham to become the first mayor of San Francisco, elected by a margin of 246 votes. Geary went home to western Pennsylvania in 1852, donating Union Square to San Francisco. His home was in Mount Pleasant, 80 miles southwest of Reading. Geary was appointed governor of the Kansas Territory in July 1856. He was a Union general at Gettysburg and then became the governor of Pennsylvania, dying in 1873, three weeks after the end of his second term. Henry Marlin Tinkcom, *John White Geary: Soldier-Statesman* (Philadelphia: University of Pennsylvania Press, 1940).

19. The 1,400-ton steamship *Sarah Sands* was built in Liverpool, England, in 1847. It sailed between that city and New York until 1849 when it was bought for use in the Gold Rush. *Sarah Sands* departed from New York on December 13, 1849, and after passing around the Horn arrived in San Francisco on June 5, 1850. The ship was overdue, because it ran out of coal. *Sarah Sands* was the second iron-hulled, screw-driven steamship in the world, with four masts as backup power. After serving the San Francisco–Panama route for a year, it returned to the Atlantic Ocean. *Sarah Sands* became the topic of a narrative by Rudyard Kipling, "The Burning of the *Sarah Sands,*" when its crew and British passengers kept it from sinking and got it to harbor. "To California on the *Sarah Sands:* Two Letters Written in 1850 by L. R. Slawson," foreword by Russell E. Bidlack, *California Historical Society Quarterly* 44, no. 3 (1965), 229–35; Rudyard Kipling, *Land and Sea Tales* (Cornwall: House of Stratus, 2009), 100–109.

20. Chinking: stuffing, or packing, to fill up the box.

21. Liverpool ware is a reference to functional tableware from the English Midlands.

22. Spencer B. Alden, twenty-seven in 1850, left a wife of less than a year to seek his fortune in California. Boarding the Pacific Mail's new SS *Carolina* in the Port of New York in January 1850, he arrived in San Francisco in early May. When he returned to Utica the next year, he was a mechanic. In 1862 he enlisted in the Union army and fought at Fredericksburg, Virginia. Alden returned to Utica to become an iron-fence maker. *Daily Alta California*, May 8, 1850; *Utica Daily Observer,* July 19, 1865; Census, 1870.

23. The steamship in question was *Mariposa*, originally to be named *Placer*, built by Walter, McKassan, and Butcher in Happy Valley, a neighborhood on the Bay in southeastern San Francisco. Intended for Stockton trade, its keel was laid on or about May 1, 1850. References to the vessel appear in the *Daily Alta California* (San Francisco), June 7, 1850, and in the *Sacramento Transcript,* June 10, 1850. Happy Valley is bounded by Market and Folsom Streets, northeasterly from Third Street to Fremont Street, which was then the Bay shoreline.

24. The "five" were probably living in Happy Valley, and they consisted of Allen and Hosea Grosh, Absolom Holland, McCaulley, and Lyons. Holland was from Union, Fayette County, Pennsylvania, born between 1801 and 1810 (Census, 1830).

25. John Hancock Gihon held a degree in medicine but never practiced. In January 1849 he moved his Philadelphia print shop to San Francisco. He had joined the Freemasons in Philadelphia, and in San Francisco he was elected as the first grand secretary of the Grand Lodge of California. He published the *Evening Picayune* briefly in San Francisco, and in 1850 he was one of eight Democratic candidates for assessor. E. A. Sherman, *50 Years of Masonry in California, 1896* (San Francisco: George Spaulding, 1898), 2:24, sec. 2. When the former mayor of San Francisco John Geary became governor of the Kansas Territory in July 1856, he named Gihon as his private secretary. Gihon went on to write *Geary and Kansas* (1857; reprint, Ann Arbor: University of Michigan Press, 2006). On January 5, 1851, Allen Grosh reports that Gihon had "gone home with his pile. He was tolerably popular here for a while but times are changed, public opinion became frail and he found it time to slope" ("slope" being a now-archaic term for "slink away").

26. Allen spells the biblical place of sin and destruction "Saddam." Gen. 19:24 and following. The reference in the letter is to the third "great fire" in San Francisco, on June 14, 1850, an event that led to fresh calls for, and then the organization of, a fire department. The fire destroyed approximately three hundred buildings in the area bounded by the Bay and Clay, California, and Kearny Streets.

27. Rockdale is 44 miles northeast of Reading, Pennsylvania. It is a farming and woodland area along the Lehigh River.

28. Sear or sere: to dry.

29. Frustrated with mining competition by Chileans, Mexicans, and Chinese, and with the British convicts from Australia, the California Legislature passed "An Act for the Better Regulation of the Mines, and the Government of Foreign Miners," April 13, 1850. Three days earlier, on April 11, another act declared that a person bringing a convict into the state faced a one-thousand-dollar fine per convict, plus three months in prison. A "Col. Gift" was then appointed collector of the "Alien Tax" in El Dorado County. Theodore H. Hittell, *General Laws of the State of California from 1850 to 1864* (San Francisco: H. H. Bancroft, 1865), 1:456; *Sacramento Transcript*, April 30, 1850. The Clear Lake, or Bloody Island, Massacre occurred May 15, 1850, on what was then an island in northern California's Clear Lake, home of a band of Pomos. Pomos from another band had been impressed and viciously abused, and many starved to death or were killed outright by two settlers. The tribe then killed the two men. A cavalry unit led by Lieutenant Nathaniel Lyon arrived to punish the Pomos for killing two whites. Finding Pomos on the island, they slaughtered an estimated sixty to one hundred of the four hundred living there, including women and children. It was the wrong band of Pomos. Max Radin, "The Stone and Kelsey 'Massacre' on the Shores of Clear Lake in 1849: The Indian Point of View," *California Historical Society Quarterly* 11, no. 3 (1932), 266–73.

30. Thomas Butler King, a native of Massachusetts, practiced law and politics in Georgia. He was eventually appointed collector of the Port of San Francisco. At the request of President Zachary Taylor, he wrote an extensive report on California (*House Executive Document 59, 1st session, 31st Congress*, ordered printed April 11, 1850). Califor-

nia had become part of the unorganized Western Territory acquired under the Treaty of Guadalupe Hidalgo, which was signed just as the Gold Rush began in 1849. Congress had not yet made provisions for organization of the region as a territory, and by default the existing Mexican law remained in effect. Based on the Napoleonic Code and little understood by men from eastern states schooled in the Americanized common law of England, settlers needed action by Congress, itself in disarray due to slavery issues. At the time of this letter, the actual content of King's report had not reached California. King's lengthy report appeared in several formats: as an appendix to Bayard Taylor, *Eldorado: Adventures in the Path of Empire* (New York: Putnam, 1859); as a freestanding document in England; and as *House Document No. 59* by Congress in 1850 (with an extra ten thousand copies). The latter text is online at the University of Michigan's "Making of America" website.

31. "General government" was the antebellum way of referring to what is now known as the federal government.

32. "Sydney convicts" refers to men who had come from Sydney, New South Wales, in Southeast Australia. For years the British transported convicts to Tasmania Island, Australia, rather than imprison them in England. After release in mainland Sydney, New South Wales, many came to California. In San Francisco they were often called "Sydney ducks." William B. Secrest, *California Desperadoes: Stories of Early California Outlaws in Their Own Words* (Clovis, CA: Quill Driver Books, 2000), 29–57.

33. Allen imaginatively spells this as "quyottes."

34. That is, a twenty-one table, referring to the card game "twenty-one."

35. Holland had probably arrived May 5, aboard the steamship *Isthmus*. The ship had been engaged in the New Orleans–Havana trade until 1849, when it was purchased by the Laws Line and brought a group around Cape Horn, arriving in San Francisco on March 5, 1849. It then began a regular run on the Pacific route between Panama and San Francisco, Captain Hitchcock commanding, generally making the one-way trip in twenty-one to twenty-three days. When traffic on that route diminished about 1854, it began a passenger and freight service between San Francisco and Portland, Oregon. E. W. Wright, *Lewis and Dryden's Marine History of the Pacific Northwest* (Portland, OR: Lewis and Dryden Printing, 1895), 83n25; *Daily Alta California*, May 9, 1850; *New York Herald*, December 28, 1848.

36. Rusk can refer either to a hard, dry biscuit or to a toasted slice of bread.

37. The letter probably reported the death of Captain Thomas S. Loeser, September 12, 1849, in Reading. He had served with an artillery company from Reading, seeing combat during the Mexican-American War in a regiment commanded by John W. Geary, recently installed as mayor of San Francisco. Colonel Banks's letter doubtless reported the death of Loeser of "an affection of the stomach and liver contracted in Mexico." Charles K. Gardner, *A Dictionary of All Officers Who Have Been Commissioned in the Army of the United States* (New York: G. P. Putnam, 1853), 188, 529; Montgomery, *Berks County* (1886), 185–86 (see chap. 1, n. 37); Pennsylvania Veterans Burial Cards, http://www .ancestry.com, accessed November 10, 2010.

38. William Henry Seward (1801–72) served as the U.S. senator from New York from

1849 to 1861, at which time he became secretary of state under President Abraham Lincoln. Senator Seward was an eloquent and forceful proponent of California statehood, which was granted on September 9, 1850, more than three months after the writing of this letter. There had been some discussion, as part of a compromise, to admit southern and northern California to the Union as separate states, one free and one proslavery.

39. William Gwin (1805–85), himself a slaveholder from Tennessee, recognized that slavery would have no future in California, and, wishing to gain control of the California Democratic Party, he did not make provisions for slavery in the state constitution. Gwin subsequently represented California in the U.S. Senate. The growing North-South division presented Gwin with a complicated dilemma, both professional and personal. Arthur Quinn, *The Rivals: William Gwin, David Broderick, and the Birth of California* (New York: Crown, 1994); "Gwin, William McKendree," in *The Biographical Directory of the United States Congress*, http://bioguide.congress.gov/scripts/biodisplay .pl?index=G000540, accessed November 27, 2010. Leonard L. Richards, author of *The California Gold Rush and the Coming of the Civil War* (New York: Random House, 2007), noted in an interview, "Despite the power of men like William Gwin, the state in 1860 was overwhelmingly northern. Men from the Deep South totaled about 6 percent of the population, and men from the border slave states totaled about 7 percent. In contrast, about 47 percent of the state's population had northern roots. While these northern men might have been indifferent to much of the political fanfare of the 1850s—the Fugitive Slave Act, the Kansas-Nebraska Act, etc.—they were not indifferent to what happened in 1861. They knew that the Confederacy fired the first shot." *Civil War Book Review*, LSU Libraries Special Collections, Winter 2007; interview of the author by Christopher Childers, http://www.cwbr.com/(Interviews), accessed January 6, 2011.

40. California's block of gold-bearing granite was a contribution to the Washington Monument. The stone as it currently appears has the following inscription: "California youngest sister of the Union brings her golden tribute to the memory of its father."

41. The Trinity River of northern California is the longest tributary of the Klamath River. Placer gold was first found in the Trinity River in 1848 by Major P. B. Reading, who then changed the name from Smith River to Trinity River. Beginning in February 1850, California newspapers carried a number of articles and advertisements about the gold placers on the Trinity.

 1851

1. Gold Lake was a settlement in Butte County, California, located 17 miles north of Bidwell's Bar, 23.5 miles northeast of Oroville. Gold Lake was shown on maps as late as 1897.

2. The Virginia rocker, or cradle, was developed in Virginia in the mid-1800s. The device consisted of an oblong box mounted on rockers. Miners placed soil in the box and then poured water into its upper end while it was placed up and down a slope and rocked like a baby's cradle. The heavy gold particles were caught by riffles, which might be no more than a piece of corduroy cloth in the bottom of the box. More expensive, a riddle-bottom cradle had a sheet of metal or wood "riddled" with holes or depressions

to catch gold particles. The cradle-and-rocker method required a larger capital outlay than did panning, but it was capable of processing more material in a shorter time, and it also permitted a division of labor, as one worker could rock, another haul debris, and yet another load the cradle. Miners in California generally found this device awkward and inefficient.

3. Deer Creek Dry Diggings was the original name of Nevada City, California.

4. All of the brothers' prospecting up to this point had been in El Dorado County.

5. Quartz veins are composed of various minerals: quartz, calcite, metal sulfides, and possibly free gold. As veins become exposed to surface weathering, the minerals oxidize to form soluble minerals, which are dissolved and carried away or form softer compounds, which remain behind but can crumble away with further weathering. What remains at the surface is a cellular mass of the weather-resistant minerals (quartz), reminiscent of a honeycomb with the voids filled with iron oxides and, in the case of the Comstock, free gold and black secondary silver sulfides. Having lost the structural support of the sulfides, the quartz crumbles, resulting in a rubbly mass of quartz chunks in a "soil" of red and yellow oxides and maybe some valuable silver minerals and gold. Prospectors call these weathered outcrops "gossans." The resulting rubbly mass can be dug out with a pick and shovel.

6. Allen is referring to the lithograph printed on the sheet he used for pages 2 and 3 of the letter. The image called "San Francisco in 1851" was by J. H. Pierce and printed by Charles E. Peregoy.

7. See chapter 2, note 25, for a discussion of Gihon, who is mentioned in 1850 as well as here, inspiring in both cases the use of the word "slope."

8. This appears to be a warning about Jonas Winchester's brother, of whom the *Overland Monthly*, referring incorrectly to Jonas as "James," stated that he "had an older brother who was an adventurer, prone to chimeras, reckless and a leech. He extracted many thousands from James, in schemes of gold hunting and town settlements, in Trinidad [on the Pacific Coast west-northwest of Redding, California] and other portions of the coast." November 1889, 493.

9. A "freshet" is a spring flood from rain or thaw. In 1850 the Schuylkill River rose more than 23 feet at Reading, flooding the lower end of Penn Avenue, the central business district, and causing some $500,000 in damage. Benjamin Tyson, the stay-at-home treasurer of the Reading California Association, was one of three men tasked with dispersing funds appropriated to assist those with losses. Montgomery, *Berks County* (1886), 432 (see chap. 1, n. 37).

10. The letter Spencer B. Alden brought west from John and Emma was mentioned in chapter 2 in the letter dated May 13, 1850.

11. "On the *que vive*" in this context means "on the alert"; more properly *qui vive*.

12. Allen uses the term "receipt" (misspelled in the manuscript as "recipt") here and elsewhere in the way that the word "recipe" functions today. Mexican chocolate is made from dark, bitter chocolate, often mixed with sugar and cinnamon. It has a grainy texture in the mouth.

13. This would be approximately two ounces.

14. Professor James Nooney was a graduate of Yale University with a B.A. in 1838 and later with an M.A. from Western Reserve College in Ohio. Professor Forrest Shepherd graduated with simultaneous B.A. degrees from Dartmouth and Yale in 1827, and then received an M.A. from Western Reserve College, where he was a resident professor of geology and agricultural chemistry. He was a native of New Hampshire. *Triennial Catalogue of Dartmouth College, 1873; Catalogue of the Officers and Graduates of Yale University, 1892;* U.S. Passport application, http://www.ancestry.com, accessed December 30, 2010.

15. During this period, the Grosh brothers were engaged in mining about 2.5 miles west-northwest of Nashville and 3 miles northwest of Enterprise, both towns on present-day California Highway 49 between El Dorado and Plymouth, in what was known as the Sugar Loaf region. The area is on the south side of a peak now called "China Mountain." Paolo Sioli, comp., *Historical Souvenir of El Dorado County, California* (Oakland, CA: Paolo Sioli, 1883), 198.

16. James M. Colson of Gardiner, Kennebec County, Maine, was born about 1814 and appears as a carpenter in the 1870 Census.

17. Three days before the date of this letter, on April 19, 1851, the *Sacramento Transcript* reported a list of newspapers selected to publish the state laws. It is not clear who approved the list, which does not include the *Monterey Gazette*. It may have been a legislative action. The lack of income from that printing work is probably what caused B. E. Holland's newspaper to fail. Unless Jonas Winchester had earlier denied the *Gazette* such a franchise, Allen may be in error blaming Winchester for the "trick." Winchester had resigned as state printer on March 30, "driven to take this step in consequence of the depreciation of the value of state warrants." *Daily Alta California,* March 30, April 12, and May 6, 1851. The next day the governor announced the appointment of J. B. Devoe as state printer. The legislature subsequently adopted a resolution asserting that the governor lacked the authority for the appointment. The office remained vacant from March 30 to May 1, 1851, when the legislature elected Eugene Casserly as state printer. Thus, there was no state printer when the list was prepared.

18. This is probably David H. Williams, a native of Vermont, forty-one in 1850, when he was a miner at Mathenas Creek, a mile south of Diamond Springs in El Dorado County (Census, 1850).

19. During the summer of 1851, some of the region's Native Americans began attacks on miners and trans-Sierra emigrants. The Miwoks, already in armed conflict in the Yosemite area, were now resisting white incursion in the Placerville foothills, while the Maidus remained friendly to miners and settlers. The first outbreak in El Dorado County was reported from Coloma in the *Daily Alta California,* May 15, 1851. Governor McDougal happened to be in Placerville at the time and remained there to raise volunteers to "chastise the Indians." After a series of skirmishes, a Native American settlement was burned on June 5. The Miwoks retired south, and relations slowly eased. See the *Sacramento Transcript,* the *Sacramento Daily Union,* and the *Daily Alta California* during this period.

20. *Ranchería* is a Spanish word for a settlement or the workers' housing area of a ranch. California newspapers of 1851 use the term to refer to long-term encampments of

Native Americans: "The Indians are reported to have a large *rancheria* some miles . . ." (*Daily Alta California,* April 9, 1851).

21. Coming from Sonora, Mexico, the mule had good reason to fear Native American attacks. The state of Sonora occupies the northwest section of mainland Mexico. For four centuries, the mountainous country offered both sanctuary and opportunity to indigenous people. Between 1831 and 1849, twenty-six mines, thirty haciendas, and ninety ranches were abandoned because of raiding Comanche bands. In 1851 Geronimo was emerging as a leader among the Chiricahua Apaches there. The Yaquis were also known to conduct raids in Sonora. Susan M. Deeds, "Indigenous Rebellions on the Northern Mexico Mission Frontier," in *Native American Resistance,* edited by Susan Schroeder (Lincoln: University of Nebraska Press, 1988), 1–29.

22. This may have been one of two Colt "Dragoon" models manufactured in some form until 1860.

23. The Albany, New York, *Argus* was published from 1813 to 1921.

24. John McDougal was governor at the time the letter was written. Nevertheless, McDougal's first message to the legislature was not delivered until January 1852. Allen's reference here is to Peter Burnett, the first governor of California, who sent his message to the legislature a few days before resigning, six months before this letter. Burnett had served only one year of his four-year term, when "he voluntarily stepped down" on January 9, 1851. He had been unable to develop a rapport with the legislature. In December 1850 he recommended adoption of the Civil Code of Louisiana, which included the French-based provision of "guilty until proven innocent." Then, in a message delivered January 7, 1851, he had suggested the exclusion of "persons of color" from California and the imposition of capital punishment for "grand larceny and robbery," specifically mentioning stealing horses and cattle. The full text of his message appears in two parts in the *Sacramento Transcript,* January 10 and 11, 1851.

25. That is, he checked to ensure the percussion caps were properly in place so that the firearms would discharge when their triggers were pulled.

26. "Sulphuret" is an obsolete form of "sulfide," any compound of sulfur and a metal.

27. J. Neely Johnson, an attorney who later became governor of California and then president of the Second Constitutional Convention in Nevada Territory, was a '49er, arriving in the gold country at the age of twenty-three. He was quoted in the late summer of 1850 as reporting gold was being obtained on the headwaters of the Carson River, in California's Alpine County, "in such quantities as to make mining profitable." *Sacramento Daily Transcript,* August 21, 1850. On June 22 the *Transcript* had suggested that "Californians have enough to do in searching for hidden treasure of their own side of the mountains, and can well afford to leave the other alone . . . [for now]." No other reference to the "great Carson River vein" has been found, but it is possible that this is what inspired Allen Grosh's comment.

28. In Gold Rush California there were no regional or federal laws to govern mining. Each locality (stream, bar, hill, etc.) was constituted a "mining district," and the miners in a public meeting resolved upon a code of laws for the district. These usually addressed the size of ground each man should be allowed to claim and the conditions for main-

taining that claim. As disagreements arose, further public meetings were held to modify or clarify the mining-district laws. Sioli, *Historical Souvenir*, 97, reprints an example of these laws.

29. The full quote is "Hope deferred maketh the heart sick: but when the desire cometh, it is a tree of life." Prov. 13:12.

 1852 AND 1853

1. Schenck Glass, a jeweler born ca. 1830, was married and lived in Grass Valley, California (Census, 1860).

2. A "coulter" is a sharpened iron rod or knife placed just ahead of the steel plough blade to break the ground in front of the blade. The blade then lifts and turns the soil. As noted in the letter, the coulter also dislodges small rocks and, by jamming against larger rocks, protects the more expensive plough blade from damage.

3. Tartar emetic refers to poisonous, odorless, transparent rhombic crystals or white powder with a metallic, sweetish taste. Medically, it was at one time used as an emetic, a counterirritant, and an expectorant, as well as to produce sweating and to treat several diseases. It frequently produced toxic side effects, however. When applied to the skin, it can cause a sore, pustular eruption that the brothers regarded as having a healthy effect. Because this substance was so poisonous, its use declined.

4. The Coloma Road was one of the primary routes to the goldfields of El Dorado County. The road started outside Sacramento at Sutter's Fort, also known as New Helvetia, then proceeded to Willow Springs near Folsom, Mormon Island, Green Valley near Rescue, and Rose Springs, then turned north at Tennessee Creek, crossing Dry Creek and Weber Creek and continuing to Coloma. "Hangtown," mentioned in the letter, was an early term for Placerville.

5. The name Glen Hannah is lost to history, but it is clear that it is in the foothills of the Sierra.

6. In the summer of 1852, Reverend Aaron B. Grosh published *The Odd Fellows Manual*, which included a letter from the grand master of the organization authorizing the publication. Reprinted many times, it remains in print today, although it is not used as closely as it was in the past. The Independent Order of Odd Fellows is a semisecret fraternal organization reminiscent of but distinct from the Free and Accepted Masons. Thomas Lewis Disharoon (sometimes Disheroon) was born in Somerset County, Maryland, ca. 1803. He settled in St. Louis, Missouri, after his wife and daughters died ca. 1840. He remarried in Missouri before traveling to California. Disharoon returned to the East in a bid to persuade Congress to validate four California land warrants. The Disharoon family records were accessed January 10, 2011, at http://www.ancestry.com. See also the *Congressional Globe*, May 23, 1856.

7. Hunkerism relates to a division within the Democratic Party Convention dating to the mid-1840s when the Hunkers (the faction in support of southern Democrats) and the Barnburners (the faction in support of abolitionism) opposed one another. Hunkerism refers to an attitude that was conservative and against any progress. The term derives

from either the idea of "hunkering down" in one's position or that of someone who "hankers" after an office; the first edition of the *Oxford English Dictionary* points to the former. See chapter 1, note 32, for a discussion of Barnburners.

8. 1 Thess. 5.

9. Warren Reinhart Grosh did not become a doctor; he became a farmer in Cecil County, northeastern Maryland, not far south of the Pennsylvania line, southeast of Lancaster.

10. The phrase more properly, and in full, is *en poco tiempo*, "in a little while."

11. *Manada* is Spanish for "herd," a group of animals that feeds and travels together. The bloodline reference is not to recognized horse breeds. At the outset of the Gold Rush, the broodmares of California were largely Spanish blood brought in via Mexico during the Mexican occupation of California. Then more horses were brought from Sydney, colony of New South Wales, Australia, which were of English origin. Eventually, American-bred horses came in with overland emigrants.

12. This may be a reference to the Shoshones, although their homeland is in the central and eastern Great Basin and extending to the north and south outside of the Great Basin. The area of western settlement of the eastern slopes of the Sierra was Washoe territory. Homelands of the Western Shoshones were generally north and east of the settled area of western Utah Territory, but close enough for trading to occur.

13. A rod is 16.5 feet.

14. The term "Bushrangers" was initially applied to convicted British criminals transported to the island colony of Tasmania who escaped and lived in the Tasmanian Island Outback as bandits. The writer is applying the term to these more dangerous, hardcase convicts among the immigrants from Australia.

15. Mormon Station was an emigrant stop on the Carson River trail at the eastern base of the Sierra Nevada. On September 20, 1855, the residents voted to change the name to today's "Genoa."

16. Lake Valley is home to modern South Lake Tahoe. It is a natural valley extending from the south shore of the lake to the mountains to the south.

17. Today spelled "Washoe Valley," this large basin is north of Eagle Valley (which is home to Carson City), itself north of Carson Valley, the location of Mormon Station. The Truckee Meadows, home today to Reno and Sparks, extends on an east-west axis to the north of Washoe Valley. To the east of Washoe Valley is the Virginia Range, a line of north-south peaks, which contain the Comstock Lode. Virginia City sprang to life on the eastern slope of the Virginia Range in 1859.

18. Today spelled "Truckee," the river was named by emigrants for a Paiute leader they called "Chief Truckee." The three lakes are Fallen Leaf, Cascade, and Tahoe. The upper Truckee River flows into Lake Tahoe at Lake Valley. Approximately one million years ago, giant glaciers inched down the Sierra block, forming Fallen Leaf and Cascade Lakes, both of which have terminal moraines separating them from Tahoe, although both send flow to Tahoe through Taylor and Cascade Creeks, the latter by way of a waterfall at Emerald Bay. Water from Lake Tahoe then flows out into the lower Truckee at the north end, eventually reaching Pyramid Lake, a terminal body of water. Michael J. Makley, *A Short History of Lake Tahoe* (Reno: University of Nevada Press, 2011).

19. Opals are chemically similar to quartz, amethyst, and glass. Unlike most gemstones, however, opals are not crystals. They are primarily composed of silica and water. An article in the *Sacramento Daily Union*, March 30, 1855, reported the following: "The Opal.—We were shown, yesterday, by Mr. Thomas Disheroon, of Mud Springs, two extremely pellucid pebbles that were recently found at Gold Lake, east of the Sierra Nevada. They are pronounced by jewelers to be opals. The end of one of them has been fractured, developing a play of colors extremely beautiful. Their external surface is otherwise rough and blurred by attrition. The one weighs one ounce and three pennyweights, and the other is one pennyweight heavier."

20. The term "cholera morbus" was used in the nineteenth century for nonepidemic cholera and for other intestinal illnesses, particularly those involving vomiting and diarrhea.

 1854 AND 1855

1. Mount Sugar Loaf, today called China Mountain, lies about 2.5 miles west-north-west of Nashville, then called Quartzville, California. The Grosh brothers worked in an area on the south side of the mountain then called Hise's Ravine, named for John Hise. It is in the area between Slate Creek and Fanny Creek. In the early days, miners in the ravine found numerous good-size nuggets, one valued at eight hundred dollars. Sioli, *Historical Souvenir,* 90, 198 (see chap. 3, n. 15).

2. They were using the leather tops of worn-out knee-high boots as material for moccasins.

3. The brothers mention the Kansas-Missouri issue on six occasions between February 7, 1855, and May 15, 1856. While the California newspapers during that time had carried news of the Kansas-Nebraska Act debate in Congress, there is little discussion in those newspapers of the growing contention between the proslavery forces, based in Missouri, and the incoming settlers in Kansas, largely from the north. The brothers wrote their last comment on the developing situation six days before the bloody attack at Lawrence, Kansas, on May 21, 1856. Based on information available at the time from California newspapers, the brothers were expressing a desire to go to Kansas to vote against slavery, not to go as warriors. The proslavery Missourians were using bullying and beatings to intimidate antislavery voters during the time the brothers wrote of it. The "Bleeding Kansas" trouble was yet to come when the brothers stopped mentioning Kansas. The first mention in California newspapers of the Lawrence incident came with the arrival of the SS *Golden Gate* mail ship in mid-June. *Daily Alta California,* various issues, 1851–54; *Sacramento Daily Union,* various issues, 1851–54; Jay Monaghan, *Civil War on the Western Border, 1854–1865* (New York: Little, Brown, 1955). See newspapers available at http://cdnc.ucr.edu/cdnc, accessed March 4, 2011.

4. Sir Walter Scott, *St. Valentine's Day; or, The Fair Maid of Perth,* first published in 1826.

5. A reference to Niccoló Machiavelli and what has been called "his Marxist moment": *The Prince,* chapter 9, "Of the Civic Principality."

6. The California Legislature held more than fifty votes between 1855 and 1857 in

an attempt to appoint someone to fill a U.S. Senate seat that remained vacant for more than a year. Eventually, William Gwin won reelection, and Joseph Broderick secured the second seat previously held by John B. Weller. While Gwin had been involved in a duel, Broderick had not by that time, although the two eventually faced one another in a duel with rifles at 30 yards. Broderick was blocking Gwin's reelection and received a mortal wound in an 1859 duel with David Terry. Quinn, *The Rivals* (see chap. 2, n. 39).

7. Governor John Bigler (1805–71), Democrat, was the third governor of California (1852–56). William Bigler, his younger brother, was governor of Pennsylvania (1852–55).

8. Johnson's Cutoff, opened in 1852, roughly follows Highway 50 from Glenbrook, Lake Tahoe, proceeding to the south and west to John C. "Cock-Eye" Johnson's ranch and station near Placerville. The cutoff was 50 miles shorter and 2,000 feet lower than the Carson River Route (Highway 88). John Winner, "Johnson's Cutoff Pit Project," *Trail Talk* (California-Nevada Chapter, Oregon-California Trails Association) (Fall 2009).

9. In referring to "a quire of fragment and scraps," Allen is employing printer's language to mean that he has some odds and ends by way of paper. A quire is generally twenty-four sheets.

10. They have left Little Sugar Loaf and returned or, as the author put it, "moved up," meaning "north," to Big Sugar Loaf, about a mile away.

11. Pauline Libhart Grosh, born ca. 1837, was the wife of Warren Grosh (Census, 1900).

12. Frances J. Hoover (ca. 1812–64) in 1850 was a trader in Marin County, California, and then in 1860, a miner at Mud Springs (Census, 1850, 1860). He left Baltimore on April 24, 1849, as one of thirty-two passengers aboard the schooner *Creole*. Marin County Genealogical Society; *Daily Alta California*, January 24, 1850. Hoover provides valuable insights regarding the silver prospecting in the area of the Comstock in his published recollection: "From Early Days: The Silver Discovery in Washoe Valley; Relic of Ancient History; How Francis J. Hoover Found and Lost Wealth—Posthumous Papers on Mining Adventure," *San Francisco Call*, May 6, 1894.

13. David Clingan, which Allen spelled "Clingham," was born in 1828 and left Baltimore aboard the *Creole* with Frances J. Hoover. He and Francis Hoover settled in Marin County, where Clingan was a stock raiser in 1860 and a trader in 1870 (Census, 1860, 1870). He served as a representative in the California Legislature of 1853 to 1854. *Spinazze's Index* (see chap. 1, n. 5); California State Library reference desk.

14. By "receipt," the writer means "recipe" (see chap. 3, n. 12). The writer is referring to the tartar emetic the brothers had previously discussed (September 9, 1852; October 11, 1852; May 18, 1855; see chap. 4, n. 3).

15. The writer probably spoiled the paper with drops or smears of ink.

16. "Threw to the deuce" is probably Allen's way to reference the devil without calling on his name. A clear use of the phrase as meaning devil or hell, "Go to the deuce," appears in the opening paragraphs of Emily Brontë's 1847 novel *Wuthering Heights*. Consider the traditional phrase and associated folk belief "Speak of the devil, and he will appear." This is reminiscent of saying "gosh" for God to avoid blaspheming, although avoiding the name of the Divinity is out of respect, while avoiding saying the name of the devil is out of fear. "Spanish fashion" mining meant working with a pick, shovel, and pan.

17. The "Friends of Freedom" was not so much an organization as a concept. From 1839 to 1858, the Boston Female Anti-Slavery Society annually published *The Liberty Bell, by Friends of Freedom*. In March 1854 a "Convention of the Friends of Freedom" was held in Boston. Two years later, in March 1856, Senator Charles Sumner of Massachusetts gave a speech titled "The Crime Against Kansas," in which he applied the expression to antislavery troubles in Kansas and the efforts being made to "drive the Friends of Freedom out of the territory," thus deterring other "friends" from coming and from controlling the Kansas government.

18. The Lone Star Association was formed in 1848 as part of an effort to seize Mexico for the United States. The movement faded in 1850–51, but it nevertheless led to the filibustering expedition of Henry Watkins and William Walker in Lower California (i.e., Baja California) and Sonora, Mexico, in 1853 and 1854. Walker led a similar attempt to capture Nicaragua beginning in 1855 and continuing until his death in Honduras in 1860. James O'Meara, "The Famous Filibuster," *California Magazine*, January–June 1858.

19. The reference to chivalry, in this context, means "proslavery."

20. "Whom the gods would destroy, they first make mad [insane]." An early appearance of the well-known quote is in the Greek play *Medea* by Euripides. Allen refers to those of the South who imagined not only secession from the Union, but also Southern expansion of a slave-based agricultural empire dominating the entire Gulf of Mexico.

21. Appearing in the letter as "Tart. of Antim.," this refers to potassium antimony tartrate, commonly called "tartar emetic," as previously discussed (chap. 4, n. 3). The text implies that A. B. Grosh had written to warn his sons against the ill effects of the ointment.

22. Early geologists tended to explain everything they could not otherwise understand as a general effect of the Great Flood as described in the Bible. Allen's dismissal of the approach suggests a progressive attitude regarding science and biblical teaching in the context of the 1850s.

23. There are several mentions of the California drought in newspapers during the summer of 1855. The *Monterey Sentinel* is quoted in the *Daily Alta California* of July 27, 1855: "The drought has in some places proved so severe as to almost entirely ruin the grain fields, and in some other parts where the drought is not so bad, rust and smut are prevalent." The *Sacramento Daily Union* of July 7 observes, "It is not quite a fortnight since a single drop of rain has fallen in [the east Bay area]." On July 23 that paper, with the headline "The Blessed Rain," reported, "There was quite a shower of rain, says the *Placer Herald*, at Auburn, Yankee Jim's, Iowa Hill and Dutch Flat, on Wednesday evening."

24. "The Wandering Arab" is a recurring phrase during the early 1800s. For example, John Quincy Adams used that as a name for Mohammed during the conflict with the Barbary pirates over America's refusal to pay tribute and ransom. The *New Monthly Magazine and Literary Journal* for the first half of 1825 carried a romantic poem, "The Lay of the Wandering Arab," and the October 1856 issue of the *Journal of Sacred Literature and Biblical Record* refers to the nomadic life with that phrase.

25. *Saba:* Spanish for "you know or understand."

26. Dr. John B. Trask, M.D. (1824–79), was born in Roxbury, Massachusetts, and was

the California state geologist from 1851 to 1856. He wrote four geological reports, including *On the Geology of the Sierra Nevada,* published at about the time this letter was written. He had been in the vicinity of Sugar Loaf Peak in the fall of 1855. *Sacramento Daily Union,* November 22, 1855, http://www.consrv.ca.gov/, accessed January 16, 2011.

27. On August 14, 1861, William Louget enlisted as a private in the First California Cavalry Regiment, which was initially charged with protecting the Overland Mail from Salt Lake City to Carson Valley. The unit eventually went to New Mexico, where Louget was mustered out in March 1866.

28. Fort Plain, New York, is on the Mohawk River, 42 miles upriver from Schenectady.

29. A felon is an infection of the palm side of a finger, typically at the fingertip pad.

30. The nativist Know-Nothing political party arose in the early 1840s, concurrent with the demise of the conservative Whig Party, largely in reaction to fears that German and Irish Catholic immigrants would overwhelm America. Originally named the American Republican Party, in 1855 it shortened the name to the American Party. Secretive in nature, members when asked about the party would respond, "I know nothing of it," leading to the nickname "Know-Nothings." After peak membership in 1854, it had faded out of existence by the end of the 1850s. See Tyler G. Anbinder, *Nativism and Slavery: The Northern Know-Nothings and the Politics of the 1850s* (Oxford: Oxford University Press, 1994).

 1856

1. *Eubanks Hydraulics* was a magazine published in the 1840s and 1850s.

2. The day prior to the writing of this letter, the *Sacramento Daily Union* reported that the House of Representatives had approved the plan of Congressman George G. Dunn (R-IN) to send a delegation to Kansas to take testimony on the election process, which had been marked by violence at the polls.

3. California state government found a home in several successive locations, and the transitions from one to the next were not graceful, with various branches of government failing to agree simultaneously on each move. Note 4, below, discusses the moves.

4. General Mariano Guadalupe Vallejo (1807–90) was born in Monterey. A member of the first session of the California state senate in 1850, he offered 156 acres of his Rancho Suscol, on which to build a capital away from the cramped quarters of San Jose. He also offered to pay for a considerable amount of the construction. The offer was accepted, and the town of Vallejo was founded at the site, along with nearby Benicia, named for his wife. The legislature convened in Vallejo in 1851, but construction problems and a soggy location led to a move to Sacramento. Madie Brown Emparan, *The Vallejos of California* (San Francisco: Gleeson Library Associates, 1968).

5. By referring to themselves as "old batches," the writer is asserting that they are still bachelors.

6. The Arabic phrase is *la ilaha illa Allah,* "There is no deity except Allah," a cornerstone of Islamic faith. But the writer probably meant to use the phrase *Insch Allah:* "the will of God" or "it's all in God's hands."

7. "Peculiar Institution" was a period euphemism for slavery.

8. Christopher Houston "Kit" Carson (1809–68) left his home in what is now Missouri at age sixteen and earned a national reputation as an explorer and guide. He helped John C. Frémont in his exploration of the West and played a role in California during the Mexican-American War (1846–48).

9. The ever-optimistic Mr. Micawber of Charles Dickens's then-recent novel *David Copperfield* (1850) would counsel that "something will turn up," making this a proverb of the day.

10. When hitched to an *arrastra,* a horse or mule drew a beam dragging heavy rocks in an endless circle. Brought to California by Mexican miners, an *arrastra* consisted of a circular patio of hard rocks, with a post in the center from which a beam extended. Heavy rocks were attached to the beam by chain or rope. Gold-bearing quartz was then strewn over the patio. As the horse, mule, or even the miner himself dragged the beam around the circle, the quartz was broken up to free the gold. Hosea describes the chemistry involved in a postscript to the January 13, 1857, letter. The name comes from the Spanish *arrastrar,* to drag, and *arrasar,* to destroy. In this letter, it is misspelled as *arastir.*

11. The idea that the pen is "mean" suggests that it is a poor one in some manner and not working well. Perhaps the metal nib is worn, causing it to snag or otherwise perform in a less than satisfactory way. The reference to making "grub" is an assertion that it is as hard to make the pen work as it is to earn one's grub (earn enough to eat) while mining.

12. Fillmorites were those favoring the election of Millard Fillmore. Fillmore was the last president (1850–53) of the United States representing the Whig Party. He had succeeded President Zachary Taylor upon the latter's death. Fillmore hoped his reluctant support of proslavery, pro-Southern policies would preserve the Union. Ultimately, Fillmore's actions placed both him and the Whigs in disfavor, hastening the decline of the party.

13. This word is represented by only ". . . ion." It may also have been "nation."

14. The San Francisco Committee of Vigilance has been the subject of numerous books and extensive research. Robert M. Senkewicz, S.J., *Vigilantes in Gold Rush San Francisco* (Stanford, CA: Stanford University Press, 1985). In 1851 the San Francisco Committee of Vigilance had imposed a sort of order on the community, but by the mid-1850s, some may have looked to that earlier period with nostalgia. James King, who fashioned himself "James King of William," wrote a series of challenging editorials in his *San Francisco Daily Bulletin* beginning in late 1855. On May 14, 1856, he wrote a scathing editorial about James P. Casey, editor of the pro-South *Sunday Times,* who then shot King that afternoon. This set the stage for the activation of the San Francisco Committee of Vigilance in 1856.

15. "Wireworking" refers to the idea of "wire pulling," that is, maneuvering or manipulating a situation for political gain.

16. Senator John B. Weller from California (1852–57), Senator William McKendree Gwin from California (1850–55 and 1857–61), John McDougal, the second governor of California (1851–52), and Congressman Preston Brooks of South Carolina were all Democrats who would not have had the support of the Grosh family. Brooks was involved in a particularly notorious incident when he struck Senator Charles Sumner (Massachu-

setts) with a cane on the floor of the U.S. Senate on May 22, 1856, as a consequence of the ongoing debate over slavery. Congressman Philemon Thomas Herbert (1825–64) served one term in the House of Representatives (1855–57) from California. A Democrat from Alabama, Herbert supported the South and was, consequently, out of touch with the majority of Californians. He later died of battle wounds while serving as a lieutenant colonel in the Confederate army. The discussion here about Herbert refers to an incident, also in 1856, during which the congressman shot and killed the headwaiter at the Willard Hotel in Washington, DC. Although Herbert was acquitted, the damage to his career was profound.

17. William L. Dayton was Frémont's running mate in the bid for the White House. Their campaign was "Free Men, Free Soil, Frémont."

18. The Wilmot Proviso, named for Pennsylvanian congressman David Wilmot, was proposed in 1846 to address the question of slavery in territories gained as a result of the Treaty of Guadalupe Hildago. The provision would have banned slavery in newly acquired lands. The proposal passed the House but failed in the Senate.

19. "Barnburners": see chapter 1, note 32.

20. Horace Greeley (1811–72) was a founder of the Liberal Republican Party and the editor of the influential *New York Tribune,* which promoted, among other things, an uncompromising abolitionist agenda.

21. Silver at the time was selling for roughly $1.60 an ounce, suggesting that a ton of the material would contain more than thirty ounces of silver. After the strikes of 1859, the better Comstock ores produced much higher values of silver.

22. Fire brick, used to line fireplaces, boilers, and so forth, is made of kaolin, a clay mined in the 1800s at Bath, South Carolina. Ceramists in Bennington, Vermont (Fenton, Farrar, and Norton), were importing kaolin clay from Bath in the 1830s and 1840s. They relocated their factory to Bath in 1856. See http://madpotter -oldcanalpottery.blogspot.com/2009/11/col-thomas-j-davies-palmetto-fire-brick.html, accessed January 10, 2011.

23. John Mawe, *A Descriptive Catalogue of Minerals Intended for the Use of Students* (1816).

24. Spelled here "scoriatious." The term refers to "scoria," an igneous, light rock infused with air.

25. The brothers were seeking one of two books by James D. Dana: *System of Mineralogy* (1837) or *Manual of Mineralogy* (1848). Dana (1813–95) was an American geologist born in Utica, New York. John Lee Comstock (1789–1858) published *Elements of Mineralogy* in 1827.

26. "Platina" apparently refers to platinum in its natural form.

27. This is a reference to Frank Antonio, discussed in the introduction.

28. Hope Valley is located along U.S. Highway 50 below Carson Pass on the old Carson River Route. Colonel William "Uncle Billy" Rogers developed a copper mine there in the 1850s. He was widely known in Carson Valley in the 1850s. Rogers came west from Danville, Illinois, in 1849 and was the first sheriff of El Dorado County. Hope Valley is on the Carson River Route, and its copper mine operated for more than thirty

years. Uncle Billy left Hope Valley in 1859, after being unable to prevent the hanging of his friend "Lucky Bill" Thorrington. He became the first settler in Ruby Valley, east of the Ruby Mountains in Elko County, Nevada Territory, as a government Indian agent and operator of the Ruby Valley Pony Express and Overland Stage stop. Sioli, *Historical Souvenir* (see chap. 3, n. 15); Mary McDougall Gordon, ed., *Overland to California with the Pioneer Line* (Stanford, CA: Stanford University Press, 1983); Zanjani, *Devils Will Reign*, 18–19, 24, 50, 74–75, 101 (see introduction, n. 10).

29. A full account of the proceedings appears in the *Sacramento Daily Union*, July 4, 1851.

30. Colonel John Reese and Enoch Reese were both born in New York, where John was a trader when A. B. Grosh was a minister there. They came west in 1849 and were merchants in Salt Lake City when a new clerk told them of Carson Valley and sold them a claim there. In the spring of 1851, arriving with ten wagonloads of goods, Colonel Reese established a trading outpost at Mormon Station. The older brother, John, became a prominent figure in the valley, taking a leadership role in organizing the early government and the creation of the Nevada Territory in 1861. Enoch, who the same year took up a claim in Spanish Fork, had joined John in Carson Valley in 1855. Milton R. Hunter, *Brigham Young the Colonizer* (Salt Lake City: Peregrine Smith, 1973), 256; Myron Angel, ed., *History of Nevada, 1881* (Oakland: Thompson and West, 1881), 38; John Reese, "Mormon Station," in *Nevada Historical Society Papers, 1913–1916*, 186–90; Zanjani, *Devils Will Reign*, 12–13, 56, 62, 72, and 120.

31. A year's subscription to Horace Greeley's *New York Weekly Tribune*, a weekly edition, was two dollars cash: see the masthead, December 6, 1856, acquired from the Pelletier Library, Allegheny College, Meadsville, Pennsylvania; information provided courtesy of reference librarian Don Vrabel.

 1857 AND 1858

1. ¿*Quien sabé?*—Who knows?

2. Henri-Victor Regnault (1810–78) authored this classic work on chemistry, appearing initially in French and translated into English in 1852. By 1856 it was in a third English edition.

3. Lixiviate: to separate into soluble and insoluble constituents by the percolations of a liquid. At the end of this paragraph, the statement indicates that more can be added to the solution when mercury is not present.

4. In 1850 George Brown lived alone in El Dorado County (Census, 1850). In 1853 he was the station keeper at what is now the site of Fort Churchill in Lyon County, Nevada. Angel, *History of Nevada*, 36 (see chap. 6, n. 30). His death is addressed in note 13.

5. The infamous *Dred Scott* decision, handed down by Judge Roger B. Taney on March 6, 1857, ruled that African Americans could not be regarded as citizens. The *Sacramento Daily Union* discussed the ruling in two separate articles, April 13 and 16, 1857.

6. Named after Frank Antonio; see the discussion in the introduction.

7. The *Sacramento Daily Union*, May 2, 1855, reported on a meeting of quartz miners at Mud Springs to adjudicate the rights of Francis J. Hoover, who asserted that on February

2, 1855, someone had jumped a claim he had held since August 13, 1851. Hosea Grosh served as secretary, and John Hise and Allen Grosh were on the eleven-man jury.

8. An *arrastra* (see chap. 6, n. 10).

9. "Dead expense": from a French idiom—*dépense sèche*—an expense that leads to no particular benefit.

10. Litharge is lead oxide (PbO).

11. Carbonate soda is sodium carbonate or "washing soda."

12. A brasque of charcoal is a paste made of powdered charcoal and some other substance, which may have included mud, molasses, or tar, and used to line fireboxes in which ore is roasted in order to cause a chemical reaction.

13. Gravelly Ford (in the letter as Gravelles Ford) was an early crossing of the Humboldt River in present-day Eureka County near Beowawe. Members of M. W. Buster's California-bound train of wagons and cattle, from Greenfield, Missouri, had begun killing Native Americans on sight at Fort Laramie. At Gravelly Ford, they killed eighteen Shoshones after the Native Americans supposedly "bantered the boys to fight." The killing of Brown and T. B. Jones and J. P. Jones occurred shortly thereafter, but it is unclear whether Buster's men or the Shoshones were guilty of the killings. Mrs. Laura M. Dettenreider maintained that she was the one who told the Grosh brothers of Brown's murder, having herself recently passed through Gravelly Ford. She also recalled seeing Hosea after he was injured. Several accounts of the incidents at Gravelly Ford were carried in California newspapers. The Buster train killing of the Shoshones is discussed in Ronald W. Walker, Richard E. Turley, and Glen M. Leonard, *Massacre at Mountain Meadows* (New York: Oxford University Press, 2006), 96. Mrs. Dettenreider's letter in Angel, *History of Nevada*, 36, 51-52; "Arrival from the Plains—Indian Attack—Eighteen Killed," *San Francisco Daily Globe*, August 21, 1857; "Later from Carson Valley," *San Francisco Daily Globe*, August 18, 1857; *Sacramento Daily Bee*, August 19, 1857; *Daily Alta California*, August 18, 1857; and *Sacramento Daily Union*, "Letter from Carson Valley," August 17, 1857.

14. One school of thought holds that Brigham Young, the head of the Church of Jesus Christ of Latter-day Saints, employed a group known as Destroying or Avenging Angels. Sometimes called the Danites, these men reputedly killed anyone Young felt needed to be punished for opposing or straying from the Mormon faith. Harold Schindler, *Orrin Porter Rockwell, Man of God, Son of Thunder* (Salt Lake City: University of Utah Press, 1966). The reference to Buster's train probably refers to a Captain Buster's wagon train, which had murdered the wife of one of the Shoshone chiefs in Wyoming. Mavis Shahrani, "Wagon Roads—1857 Season," in *Frederick West Lander: A Biographical Sketch*, edited by Joy Leland, Publications in the Social Sciences no. 15 (Reno: Desert Research Institute, 1993), 82.

15. Two sources identify Captain W. B. Galphin, "Capt. Galvin," as a "partner" of the Grosh brothers: Mrs. Laura M. Dettenreider and William Dolman. Dolman describes the shooting incident in detail and mentions Galvin a number of times. On January 30, 1858, at Johntown, Galphin provided the first signature on the "Bylaws of Quartz Miners." The other eleven names included James Finney, Emmanuel Penrod, and Dolman. Angel, *History of Nevada*, 51; Austin E. Hutcheson, "Before the Comstock, 1857-1858: Memoirs of William Hickman Dolman," *New Mexico Historical Review* (July 1947) (reprinted by the

University of Nevada, Reno), 23–25; Marion Ellison, *An Inventory and Index to the Records of Carson County, Utah and Nevada Territories, 1855–1861* (Minden, NV: Grace Dangberg Foundation, 1984), 77.

16. The text presented here is augmented by the transcription of the letter by A. B. Grosh, some of which appeared in the obituary for Hosea B. Grosh that appeared in the *Christian Ambassador* (Auburn, NY), November 14, 1857.

17. 1 Cor. 15:55.

18. Dr. Benjamin L. King was one of only two physicians in western Utah Territory in 1852 and one of only four residents of Eagle Valley. Born in New York State in 1802, he settled at the mouth of a canyon in Eagle Valley. In 1852, when the canyon became the beginning of the Johnson's Cutoff to California, King opened a station for emigrants on his claim. The canyon then became Kings Canyon. Johnson had done little to improve that portion of his trail to California, and it was used by few emigrants after the first year. With a brewery on the property, the King station became a place of public resort. King became active in local matters, running unsuccessfully for Carson County sheriff in 1857, and he was active in efforts that led to the creation of the Nevada Territory in 1861. Angel, *History of Nevada*, 33–34, 42–43, 531; *Sacramento Daily Union*, August 24, 1857; *Nevada Historical Society Papers, 1911–12*, 185.

19. Jacob H. Rose, born ca. 1812 in New York, was one of the first settlers of Washoe Valley. The name appears in 1855 in the first court case in Carson County, Utah Territory. He later lived as a farmer in the Beowawe, Nevada, area, in modern-day northern Eureka County. Myra Sauer Ratay, *Pioneers of the Ponderosa* (Sparks, NV: Western Printing and Publishing, 1973), 49; Census, 1870.

20. Dyspepsia: upper abdominal bloating, tenderness, and pain.

21. Dr. Charles D. Daggett arrived in the Carson Valley in late 1851. He rented a log structure with a man named Gay and took an active role as district attorney and county assessor in Carson County, Utah Territory. When the Mormons left in the summer of 1857, Daggett supported the move to have a separate territorial status for the western part of the Utah Territory. For many years, Daggett lived in a cabin on the creek at the mouth of old Kingsbury Grade in Carson Valley, at the base of a rough Washoe trail to Lake Bigler, as Tahoe was known at the time. Angel, *History of Nevada*, 31, 41–42, 333; *Nevada Historical Society Papers, 1911–12*, 200. Daggett studied at the Berkshire Medical College in Pittsfield, Vermont, and was also accredited as a lawyer by Utah Territory (Nevada State Archives).

22. The composition of Blue Mass, a common medication in the nineteenth century, depended on the person who was mixing the ingredients, but they all generally included mercury, making the medicine toxic.

23. "Campbellite" refers to a Christian reform movement of the early nineteenth century. Thomas Campbell and his son, Alexander, were natives of Scotland living in a Scots-Irish community in southwestern Pennsylvania. They became part of a religious movement that led to the opening of two fundamentalist movements, the Church of Christ and the Disciples of Christ. Charles H. Lippy, *Bibliography of Religion in the South* (Macon, GA: Mercer University Press, 1985).

24. The word "Patience" is illegible on a damaged portion of the original letter but

is included in the transcription published by Lord, *Comstock Mining and Miners*, 28 (see introduction, n. 15).

25. William Dolman describes this incident in detail. Hutcheson, "Before the Comstock," 19-21. Dolman does not identify the deceased as an Irishman, but he does recall that the African American beat Peter O'Riley, "the big Irishman, who is often mentioned in the Comstock Lode discovery." O'Riley survived the attack.

26. John McMarlin was shot at Slippery Ford Hill, California (Kyburz on U.S. 50). His brother, James, was justice of the peace for Gold Canyon. Williams was, as the letter notes, constable for Carson County, Utah Territory. James McMarlin and Williams both took their elected offices in late 1855. Angel, *History of Nevada*, 38, 147; *Sacramento Daily Union*, September 8, 9, 10, and 14, 1857; Robert E. Ellison, *Territorial Lawmen of Nevada* (Minden, NV: Hot Springs Mountain Press, 1999), 1:79-85.

27. Brigham Young ordered the Mormons to return to Salt Lake City in 1857 in anticipation of what Young regarded as an invasion of federal troops, who were marching to Utah to secure the territory on behalf of the federal government. Zanjani, *Devils Will Reign*, 77-78 (see introduction, n. 10); *New York Times*, October 17, 1857.

28. The reference to the "medical maltreatment" of Allen's mother is without any other documented context.

29. The panic of 1857 began in September as a reaction to a recession of the financial markets.

30. Charles A. Washburn, born in 1825, was a member of the Credentials Committee of the convention.

31. Marietta and Irishtown are neighboring communities in Lancaster County, Pennsylvania.

32. On the southeastern shore of Lake Ontario.

33. "Charley" is cousin Charles R. Grosh, b. 1843, a son of Reverend Grosh's brother Charles C. Pinkney Grosh. Blaine is Alexander Blaine Grosh, 1846-1920, son of Dr. Benjamin Franklin Grosh, another brother of Reverend Grosh. Both Charley and Blaine were thus cousins of Allen and Hosea (http://www.ancestry.com, accessed February 20, 2011).

34. Reverend Grosh's brother Dr. Benjamin Franklin Grosh (1818-57) was a physician in Perry County, where he also owned a farm. He is referred to as "Dr." and "Franklin."

35. "Aunt Lizzie" is Elizabeth A. Grosh, b. 1825, wife of C. C. Pinkney Grosh.

36. "Grandfather," also referred to as "father" in letters by Reverend Grosh, is Jacob Grosh (1776-1860). The introduction includes a discussion of this individual.

37. The *Christian Ambassador*: see introduction, note 6.

38. H. H. Hain, *History of Perry County, Pennsylvania* (Harrisburg, PA: Hain-Moore, 1922), 538, includes the following that may apply to this reference: "On November 6, 1858, Mary Barton (colored) died in Buffalo Township, aged sixteen years, the newspaper accounts of her death stating that 'she was formerly employed in the family of Dr. [B. F.] Grosh as a bound girl.'"

39. Jer. 10:13.

40. The text presented here is augmented by the transcription of the letter by A. B.

Grosh, some of which appeared in the obituary for Allen Grosh that appeared in the *Christian Ambassador*, February 20, 1858. In the obituary, A. B. Grosh notes that the letter was "written on a half sheet of paper, in trembling and confused characters, unlike his usual clear and decided penmanship." Inspection of the original manuscript letter confirms this.

41. The intent here is to say that he is of noble blood and of high-minded, reputable character.

42. Richard Maurice Bucke, often called Maurice Bucke (not Morris or Buck), was born in 1837 in Methwold, England, the seventh child of Reverend Horatio Walpole Bucke (a parish curate) and his wife, Clarissa Andrews. With his parents, he emigrated to Canada when he was a year old, the family settling near London, Ontario. Richard Maurice Bucke, MD, *Cosmic Consciousness*, 16th ed. (New York: E. P. Dutton, 1951; 1st ed., 1901). "Canada West," which appears in the original letter as "C. W.," is the region above the eastern Great Lakes as it was known in the first half of the nineteenth century.

43. This appears to be a confused description of the incident related by Allen Grosh about his lost jackass in Washoe Valley.

44. Reverend Grosh then notes in the obituary that "we omit the remainder, as Allen's letter belongs here." After quoting Allen Grosh's letter, the obituary continues with some of the Harrison text. It appears that this partial text of Harrison's letter to Reverend Grosh is all that has survived.

45. Reverend Grosh added here in brackets "And doubtless grief and loneliness since the death of Hosea, did its share of the work."

46. Brother "of the mystic tie" indicates Harrison is a member of the Odd Fellows.

47. W. J. Harrison was elected as an assemblyman from Placer County in 1860.

48. "'Ask what I shall do for you before I am taken from you.' And Elisha said, 'Please, let a double portion of your spirit be upon me.'" 2 Kings 2:9–13. Elijah's mantle served as proof of the miracle of Elijah's ascent to heaven, having left his mantle behind. His son Elisha took up the mantle. To take up his mantle is to assume his role, sacrifice, and commitment as a prophet. 1 Kings 19:19–21.

49. The text of the article appears with the Grosh papers at the Nevada Historical Society, but it contains no significant details.

50. Last Chance is located on the ridge south of the main branch of the North Fork of the Middle Fork of the American River, at roughly 5,000 feet above sea level. Gold was discovered there in the spring of 1850, and the settlement dates to 1852. It is now an uninhabited ghost town in a heavily forested area.

APPENDIX A: THE READING CALIFORNIA ASSOCIATION

1. The Jones Scrapbook (Reading, CA: Berks County Historical Society Library, n.d. [ca. 1887]), microfilm. The unnamed writer of the article names the eleven Reading men who joined the Gordon company and briefly describes their difficult crossing of Nicaragua.

2. Ibid.

3. Haskins, *The Argonauts of California* and *Spinazze's Index* (for both, see chap. 1, n. 5);

George R. Stewart, *The California Trail: An Epic with Many Heroes* (New York: McGraw-Hill, 1862), chap. "49."

4. The Jones Scrapbook.

5. "The Gold Seekers," *Reading Eagle*, May 28, 1893. Andrew Sallade, Esq., forwarded to the Grosh family an early letter by Thomas Taylor dated September 24, 1849, which is now part of the Grosh Collection at the Nevada Historical Society in Reno.

6. The Taylor and Martin deaths are both mentioned in the Grosh brothers' letters.

7. Although the Census indicates that Charles was twenty-one in August 1850, he swore he was twenty-one on February 24, 1849, when he applied for a passport. More than a year had passed, so he must have been twenty-two by August 1850. Ages for the various members of the Reading California Association in 1849 presented throughout this discussion are based on passport applications (http://www.ancestry.com, accessed February 28, 2011). Taylor family genealogical information: Walter H. Ireland, "Descendants of Elizabeth Arton Boone," *Boone Bulletin* (American Order of Pioneers and Boone Family Association, Washington, DC) 2 (1931): 7.

8. Montgomery, *Berks County* (1909), 767 (see chap. 1, n. 37).

9. Land Patent Records, Bureau of Land Management State Office, Sacramento.

10. *Book of Biographies: Berks County, PA* (Buffalo, NY: Biographical Publishing, 1898), 61–62; passport applications, http://www.ancestry.com, accessed January 17, 2011; Jones Scrapbook; Census, various years.

11. "The Gold Seekers." Flack applied for a passport on February 24, 1849, providing information on his birth. His death was reported in the *New York Times*, May 27, 1884.

12. Klapp's problems with the U.S. Mint were reported in the *New York Times*, August 30, 1852. The federal court files in Philadelphia indicate no charges were filed in the matter (conversation with James Hamilton, U.S. District Court Clerk's Office, Philadelphia, February 28, 2011). He died on January 25, 1894.

13. Census, 1860, 1880, 1900.

14. Census, 1860.

15. "The Gold Seekers."

16. *Sacramento Daily Union*, March 16, 1863; *Kirshtein Family Tree*, http://www.ancestry.com, accessed March 2, 2011; Patent Records, Bureau of Land Management, Sacramento; Census, various years.

17. M. W. Wood, *History of Alameda County* (Oakland: Pacific Press, 1883), various pages; *San Francisco Call*, April 26, 1908; Census, 1900.

18. Jones Scrapbook; Zerbe's diary.

19. Haskins, *The Argonauts of California*, 487.

20. Jones Scrapbook. Spelling of some names has been corrected.

APPENDIX B: THE RECOLLECTION OF EDWIN A. SHERMAN

1. "Sherman Was There," 267 (see chap. 1, n. 7). E. A. Sherman (1829–1914) was a prolific writer, his works including his two-volume history of the Masonic Order in California. In 1861 he opened a newspaper in the mining camp of Aurora, then thought to be in California. In 1863 he was elected Nevada state controller on the ballot on which

the first state constitution was defeated. Although he was elected, there was no office to assume. He remained in Nevada until 1883, living in several locations.

2. Peter E. Palmquist and Thomas R. Kailbourn, *Pioneer Photographers of the Far West: A Biographical Dictionary, 1840–1865* (Stanford, CA: Stanford University Press, 2000), 85–86; Palmquist and Kailbourn, *Pioneer Photographers from the Mississippi to the Continental Divide: A Biographical Dictionary, 1839–1865* (Stanford, CA: Stanford University Press, 2005). In all cases, Palmquist and Kailbourn cite Sherman as their source. Correspondence from Thomas R. Kailbourn, April 5, 2011, to Robert E. Stewart clarifies the process of evaluating the Grosh brothers' alleged involvement with photography.

3. "Sherman Was There," 267.

4. An article from *Lloyds Weekly Newspaper* (London), "More News from the Gold Country," February 25, 1849, states: "January 7: The mail steamer, *Falcon*, left New York for Chagres. . . . On the same day the brig, *Thomas Walter*, left [Philadelphia] with fifty passengers, including the 'Camargo Company' for Tampico, and thence overland to California."

5. According to the *New York Herald*, March 1, 1849, "The schooner *Newton*, Capt. Smith, . . . sailed yesterday afternoon from Philadelphia, with the following passengers, destined for California. They proceed by the way of Tampico."

6. Antrim's journals and watercolor images of Mexico are in the Benajah Jay Antrim Journals, vol. 1, Manuscripts Division, Library of Congress; Census, 1850. In 1863–64 Antrim brought his equipment to Nevada, operating in Dayton and Washoe City.

7. Woods, *Sixteen Months* (see chap. 1, n. 7). Allen spells this "Guadalaxara."

8. "From Mexico. Tampico, April 1, 1849. The disturbed state of this ill-fated country is to be lamented. The disbanded officers from the Mexican army, deserters from the Americans, and a party of Indians who have been in the mountains of the Sierra Madre for some months past, marched against Rio Verde on the night of the 9th of March. They encountered the Mexican troops stationed at that town, and soon put them to flight, and afterwards committed all kinds of excesses and depredations. The troops stationed at this place [Tampico] have marched against them, under the command of Gen. de la Vega; but La Vega is too weak to attack them [300 kilometers, about a ten-day military march]. This movement has paralyzed all our commercial relations with the interior, whilst our northern neighbors are sending American goods across the country from Texas to Matamoros; and should the town of Brownsville become a port of entry, all the principal merchants of Tampico will break up here and establish houses there." *Tioga Eagle* (Wellsborough, PA), May 9, 1849.

9. Hoover, "From Early Days" (see chap. 5, n. 12); "The Real Discoverers of Silver in Nevada; Chapter from Pioneer History Concerning Two Prospectors Who Garnered Not the Fruits of Their Find," *San Francisco Call*, August 16, 1896.

10. ". . . la fotografía llegó a nuestros territorios hacia 1870; por ello, no creo posible el registro fotográfico al que Usted alude." (. . . photography came to our territories about 1870; for that reason, I do not believe possible the photograph to which you allude.) Rafael Morales Bocardo, director, Historical Archive of San Luis Potosí, e-mail to Robert Stewart, April 7, 2011.

APPENDIX C: FINANCIAL MATTERS AND THE SUBSEQUENT LAWSUIT

1. A final financial matter to address after the deaths of the brothers was the issue of the fifty dollars that they had borrowed from John Hise with the understanding that their father would repay the amount if they were not able to do so. The letters preserved by the family include a document that addressed the need to repay the amount, but the logistics of transferring the funds was certainly a challenge. How this matter was resolved is unclear.

2. M. Ellison, *Records of Carson County* (see chap. 7, n. 15).

3. The Bancroft Library at the University of California, Berkeley, has a collection of Grosh-related documents, including letters between Aaron B. Grosh and Richard Maurice Bucke (http://oskicat.berkeley.edu/record=b11219387~S1, accessed November 18, 2011).

4. Lord, *Comstock Mining and Miners,* 133 (see introduction, n. 15).

5. Ibid.

6. For Mrs. Dettenreider's interview, see Angel, *History of Nevada,* 51–52 (see chap. 6, n. 30). Reverend Grosh's statement appears on pages 52–53.

7. Francis J. Hoover provides a letter that includes information in this context: *San Francisco Alta California,* March 16, 1864.

8. Ibid., 51–52.

9. Some of the details of Mrs. Dettenreider's tale are echoed in a letter from Caroline B. Winslow, written in Virginia City to A. B. Grosh and his wife on August 19, 1879, part of the Grosh Collection at the Nevada Historical Society. Winslow spoke with Dettenreider, who described the situation in the same terms as those that appeared in her statement. Ibid.

10. Ibid., 52–53.

11. Ibid., 53.

12. For a discussion of the folklore and erroneous history that have swirled around Henry Comstock, see Ronald M. James, *The Roar and the Silence: A History of Virginia City and the Comstock Lode* (Reno: University of Nevada Press, 1998), 1–20.

13. Sam P. Davis, *The History of Nevada* (Los Angeles: Elms, 1913), 389–90.

14. Zanjani, *Devils Will Reign,* 41–46 (see introduction, n. 10).

INDEX

Abbott, William Thomas, 15, 22, 49, 60, 193; illness, 23, 24, 26; mining, 28-29, 32, 37, 40, 50, 51-52; returns to Pennsylvania, 84-85

Adelaide (ship), 72

African Americans, 173-74, 180, 234n25, 235n38

Albright, Moses, 195

alcohol, 16, 26, 47, 100, 120, 127, 128, 132

Alden, Spencer B., 63, 70, 77, 216n22

Alien Tax, 65

ambrotype. *See* photography

American Indians. *See* Native Americans

American Ravine, 202-4

American Republican Party. *See* politics: Republican Party

American River, California, 78, 189

Antonio, Frank, 11, 147, 203, 323n6

Antrim, Benajah Jay, 198

Armor, Dr., 179

arrastra. *See* Mexican milling process

Australian Americans, 66, 105, 218n32, 224n14

Avenging Angels. *See* Destroying Angels

Axe, Reuben, 15, 29, 40, 47, 50, 51, 60, 192, 194

Ballou, Reverend Hosea, 3, 9

Banks, Ephraim N., 62, 71, 216n17

Barnes, H. M., 190

Behne, John H., 23, 26, 32, 132, 210n16

Bigler, William, 112, 226n7

Bitting, Franklin, 50, 196

Blake, Ed, 53, 214n7

Blue Mass. *See* health issues: cures

Bowman, William G., 50, 196

Briner, Lewis, 40, 212n37

Broderick, Joseph, 226n6

Brooks, Preston, 142, 230n16

Brown, George, 156, 158, 163, 166-68, 173-74, 178, 198, 231n4, 232n13

Bryant, J. J., 59

Buchanan, James, 141, 150, 161

Bucke, Maurice (Morris) R., 4-5, 182-83, 184, 210n15, 235n42

Bull, Alpheus, 25, 190, 211n21

Burnett, Peter, 70, 85, 222n24

Burns, Al, 183

Buster, M. W., 166-67, 232n13

Butler, Benjamin F., 205

Butler, John M., 159

Byerley, Daniel, 40

California Quartz Miner's Convention, 147

California statehood, 65-66

Camargo company, 197-99

Campbellite, 172, 234n23

Carson, Kit (Christopher Houston), 137, 229n8

Carson River, 87, 107, 108, 222n27

Carson Valley, 102-4, 106-7, 108, 144, 147-49, 154, 171, 174, 186-87, 203

Cheesman, Thomas W., 57, 74-75

Chilean Americans, 55, 65-66, 182, 215n10

chocolate, 38, 60, 63, 77, 220n12

Christian Ambassador, 180, 190, 207n6, 235n40

Church of Jesus Christ of Latter-day Saints, 144, 148, 160; and Native Americans, 166, 184; political power, 153-55, 161, 167, 232n14; polygamy, 150, 154-55; response to threat of federal invasion, 174, 181, 234n27

Civil War, 112, 227n20